Cellular Digital Packet Data

The Artech House Mobile Communications Series

John Walker, Series Editor

For a complete listing of *The Artech House Telecommunications Library,* turn to the back of this book.

CELLULAR DIGITAL PACKET DATA

Muthuthamby Sreetharan
Rajiv Kumar

Artech House
Boston • London

Library of Congress Cataloging-in-Publication Data
Sreetharan, Muthuthamby
    Cellular digital packet data / Muthuthamby Sreetharan and Rajiv Kumar.
        p.   cm.
    Includes bibliographical references and index.
    ISBN 0-89006-709-0 (alk. paper)
    1. Wireless communication systems.   2. Cellular radio.   3. Radio—Packet
transmission.   I. Kumar, Rajiv.   II. Title.
    TK5103.2.S79   1996
    004.6'6—dc20   96-13238
                                                            CIP

British Library Cataloguing in Publication Data
Sreetharan, Muthuthamby
    Cellular digital packet data
    1. Mobile communications systems 2. Cellular Radio 3. Wireless
communication systems
    I. Title II. Kumar, Rajiv
    621.3'8456

    ISBN 0-89006-709-0

Cover design by Kara Munroe-Brown

© 1996 ARTECH HOUSE, INC.
685 Canton Street
Norwood, MA 02062

International Standard Book Number: 0-89006-709-0
Library of Congress Catalog Card Number: 96-13238

10 9 8 7 6 5 4 3 2 1

To my mother, who perished during the 1987 Sri Lankan army's onslaught on Jaffna; to my father, who taught me, by example, the important values in life; and to young Tamil men and women who gave their lives defending the freedom and dignity of the people of Thamil Eelam.

— M. Sreetharan

To my wife, Amy, and daughter, Sonali, without whose support I would not have known about CDPD.

— Rajiv Kumar

▼▼▼

CONTENTS

▼▼▼

PREFACE

Cellular digital packet data (CDPD) is emerging as a popular wireless technology for mobile data communications. This book is intended to serve as a detailed reference to CDPD.

Built as an overlay on the existing AMPS (advanced mobile phone system) infrastructure, CDPD uses the spare channel capacity of the AMPS network. In the crowding field of wireless data, whether CDPD will become the dominant data platform will depend on the availability of low-cost modems, the availability of third-party application software, and the availability of nationwide CDPD coverage. We have provided a broad technical overview of other wireless technologies and nontechnical aspects of CDPD in addition to indepth technical details of CDPD to assist the readers in understanding the role of CDPD in the wireless data arena.

The first two chapters provide an introduction to CDPD and present a comprehensive overview of existing alternate technologies that support wireless data communications. In addition, they detail key technical aspects of cellular systems that are providing the basis for the existence of CDPD networks. The third chapter describes the architectural concepts of CDPD, detailing the major subsystems within CDPD: mobile-end systems (M-ES), mobile data-intermediate systems (MD-IS), and the mobile data base station (MDBS). These three chapters will be particularly useful for those who need to understand the overall architecture of the CDPD network and are not necessarily interested in more indepth technical details. These chapters also explain the techniques behind how the CDPD network is integrated with cellular voice systems and with the existing infrastructure of existing terrestrial data networks.

The next five chapters provide detailed technical descriptions of the main components of the CDPD network: airlink interface, link layer protocol, SNDCP functions, radio resource management, and mobility management. The technical information in these chapters is based on the CDPD Specifications Version 1.1, released by the CDPD Forum. The indepth nature of these chapters will appeal to the more serious readers. Engineers, managers, and sales and marketing staff from the CDPD application development sector and the infrastructure equipment development sector will find useful technical information they need related to CDPD. Chapter 9 is devoted to network management, and details related to configuration, security and performance, and accounting management of the CDPD network are described.

The last three chapters contain information that relates to the experience gained during deploying, operating, and tuning the network. Interoperability issues, performance issues, and the impact of the different configurable parameters in tuning the network are described in detail. Important CDPD issues that govern its future growth, including the impact of the evolving voice technology, are detailed in the last chapter.

▼▼▼

ACKNOWLEDGMENTS

We'd like to thank everyone involved in the CDPD program at Hughes Network Systems (HNS), Germantown facility; especially Ashok Mehta (assistant vice president, engineering), who gave us the opportunity to participate in the CDPD program; and Jake Macleod (assistant vice president, marketing), for giving his blessings for the book. Our special thanks to all the colleagues at HNS and Hughes Software Systems (HSS) who worked together to get a successful CDPD system working and to Yoke Kung, Nick Sampson, and Hans Bhatia for architecting a successful CDPD program.

We would like to thank the CDPD Forum, Peter Wiltjer, and Roxanne Sakala for giving us the consent to use material and figures from the CDPD Specification 1.1. The core material and figures for Chapters 3 to 9 are based on this specification. Ira Brodsky's book *Wireless: The Revolution in Personal Telecommunications* was used as a key material resource for the background information for this book.

Special thanks are due to Dr. Victor Moore (IBM), who gave us the background to the CDPD history; to A. Ramanathan, who contributed the accounting section in Chapter 9; and to Douglas Conner, who created most of the artwork for the book (with special care and attention to Figure 2.14).

CDPD network managers and engineers at different CDPD sites who we have worked with gave us the insight and exposure to the practical aspects of operating a wireless network. Specifically, Joe Cruz, Scott Carlson, and Pat Walsh (Ameritech Cellular Systems); Norbert Fey and Jack Klenik (Bell Atlantic Mobile Systems)—they gave us the freedom to learn more about CDPD in their network. GTE Mobilnet stretched the network to its limits during their initial deployment days and this allowed us to really understand fault management and performance issues.

We wish to thank our publishers, Artech House; Editorial Assistant, Kimberly Collignon, who efficiently coaxed us into completing the manuscript almost on schedule; Managing Editor Laura Esterman and Art Director Kara Munroe-Brown, whose tireless efforts improved the quality of the written text and artwork; and the reviewer, whose comments had significant impact in the organization and technical content of the book. However, both of us are responsible for all remaining errors.

I'd like to thank all the colleagues at PCSI who added the subscriber perspective to all that I could write. Rusty Cashman and Sanjay Bapat of PCSI, who contributed by reviewing individual chapters. Sheldon Gilbert for allowing me time to work on this book and giving words of advice and encouragement. (R.K.)

I'd like to thank my mentor, Professor J. A. Gunawardene, who helped to shape my interest in this field. To my talented friends at PCI—Ed Crane, Nick Makris, Peter Springston, Amy Brongo, Chris Michael, Steve Vice, Naresh Venugopal, Oscar Somerlock, and Kenny Cenecius—who have been a rich source of motivation and support, and to Feizal Mohideen, who entrusted us with completing several previous technical challenges, I am grateful. Finally, I'd like to thank my sons Rajeev, Pratheev, and Sanjeev, and my wife, Mathini, for gracefully ignoring me during the period when I spent most of my spare time working on the book. (M.S.)

M. Sreetharan
Rajiv Kumar
June 1996

CHAPTER 1
▼▼▼

INTRODUCTION

Cellular digital packet data (CDPD) is a wireless technology that provides packet-switched data transfer service using the radio equipment and spectrum available in the existing analog mobile phone system (AMPS)–based analog cellular networks. AMPS technology is used mainly in the United States and therefore initial deployment and use of CDPD will predominantly be in the U.S. markets. Demand for wireless data transmission capabilities is rising sharply, and currently available wireless data technologies have not been popular with the wireless users because of their coverage, cost, and throughput (speed) limitations. Overlaid on the widely deployed AMPS radio infrastructure in the United States, CDPD has the potential to provide users with nationwide coverage that cannot be met by other competing wireless technologies. Further, CDPD promises a low cost service to its subscribers because:

- CDPD shares the use of the AMPS radio equipment on the cell sites;
- CDPD uses excess capacity in the allocated spectrum for the analog voice systems;
- CDPD network usage cost is based on the volume of data transferred and not on the connection time.

The protocol architecture of CDPD is chosen so that the available infrastructure of the terrestrial data network can be integrated as an extension to the CDPD network, as shown in Figure 1.1. In effect, the CDPD network provides a connectionless routing framework—an Internet protocol (IP) or connectionless network proto-

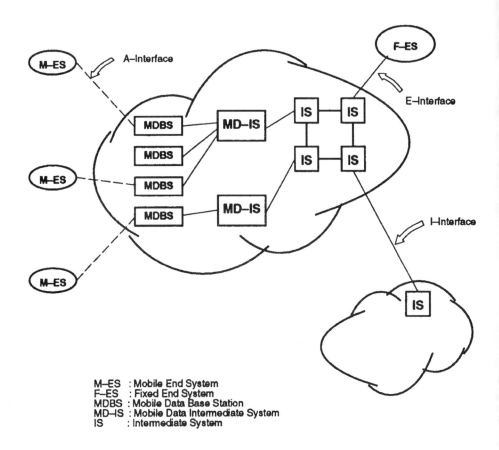

M–ES   : Mobile End System
F–ES   : Fixed End System
MDBS   : Mobile Data Base Station
MD–IS : Mobile Data Intermediate System
IS   : Intermediate System

FIGURE 1.1   Overall CDPD network architecture. (*Source:* [1].)

col (CLNP) network layer packet transfer facility—to allow a mobile end system to connect to another end system via the IP-based Internet or CLNP-based OSI networks.

This chapter provides a brief historical background to the cellular networks that serve as the basic framework for the deployment and the operation of the CDPD networks. The architecture and operation of the CDPD networks are also introduced so that the core differences between competing or complementing wireless data technologies can be identified for engineering or managerial decision making.

1.1   BACKGROUND

Within the last 20 years the wireless field has experienced unprecedented growth into a new industry. Fueled by advances in radiofrequency, satellite, and microelectronic technologies, and aided by the convenience of instant and tetherless access to telephony and messaging portable devices, wireless technology will spawn mass markets for wireless communication devices and applications.

Table 1.1 lists the chronology of landmark events in RF spectrum allocation and in the growth of the cellular technology reveals that most of the advances are relatively recent.

TABLE 1.1
Landmark Events in the Cellular/Wireless Field

Year	Event
1925	2 MHz was believed to be the highest usable frequency and the National Association of Broadcasters warn of impending frequency shortage.
1934	FCC was set up as an independent regulatory body to allocate spectrum, to define rules for services, to provide licenses to users, to certify wireless products, and to police the radio spectrum.
1958	Bell System proposal to the FCC for a 75-MHz spectrum in the 800-MHz band for use in cellular.
1970	FCC tentatively allocates 75 MHz for a wireline common carrier.
1971	Cellular system design with cells, frequency reuse schemes submitted by AT&T Bell Laboratories.
1974	FCC allocates 40-MHz spectrum with 666 channel pairs with one cellular system per market.
1978	First cellular trials conducted in Chicago.
1981	FCC implements the two-carrier per geographic area licensing.
1982	First meeting of the Groupe Special Mobile (GSM) group.
1985	Personal communication services concept emerges as a future technology.
1986	FCC allocates additional 5 MHz for each band, allowing 83 additional channels per band (A and B).
1991	CDPD patent was filed by three IBM staff members.
1993	Initial specification of CDPD by the consortium of carriers and IBM.
1994	Number of wireless users in U.S. reaches 20 million.
1995	Narrowband and broadband PCS spectrum auctioned by the FCC with the promise for near universal access to messaging, mobile telephony, and data exchange. Beginning of extensive deployment of CDPD infrastructure, third-party CDPD modem development, and third-party application software development.

Although the United States serves as the main market for the AMPS technology (over 85% of the worldwide AMPS subscribers are in the U.S.), other countries such as Australia, New Zealand, several Asian countries (such as Indonesia, the Philippines, Hong Kong, and Singapore), and several South American countries (such as Chile and El Salvador) also use AMPS as their primary wireless cellular technology [2]. CDPD, within its current stage of development, can be easily deployed in these foreign markets once the technical success and the cost advantage to the CDPD customer are demonstrated in the U.S. markets.

In 1993, anticipating the growth in the demand for wireless access to exchange data, a consortium consisting of the following carriers was formed to generate a specification for the CDPD technology:

- Ameritech Mobile Communications, Inc.;
- Bell Atlantic Mobile Systems Contel Cellular, Inc.;
- GTE Mobile Communications, Inc.;
- McCaw Cellular Communications, Inc.;
- NYNEX Mobile Communications, Inc.;
- PacTel Cellular;
- Southwestern Bell Mobile Systems;
- U S West Cellular.

The CDPD specifications embody the following major design goals:

- Compatibility with existing data networks;
- Ability to support present and future data network services and facilities through standardized access to CDPD network;
- Allowance of maximum use of the existing commercial data network infrastructure and provision of support to multiple data network protocols.

In addition, more generic goals associated with mobile systems, which include seamless roaming, security/privacy, protection of the network from fraudulent users, and support of a wide range of mobile stations (such as portable, low-power handheld) are also part of the design goals of the CDPD network.

1.2 CELLULAR DIGITAL PACKET DATA (CDPD)

The original idea for using the spare bandwidth in AMPS system to transmit data is credited to Robert Miller, Victor Moore, and Thomas Pate (IBM, Florida) and described in a U.S. patent filed in 1991. The abstract of the patent, which contains Figure 1.2, is given below:

A method for performing Cellular Data Networking (CDN) in an Advanced Mobile Telephone System (AMPS), wherein said AMPS includes a set of cellular telephone voice transceivers, each tuned to one of a pre-

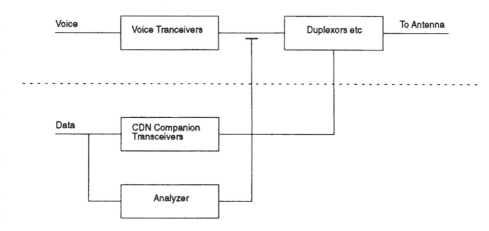

FIGURE 1.2 Organization of transceivers. (*Source:* [3].)

selected set of communication channels, and means for coupling each transceiver in said set of transceivers to an antenna to facilitate the performance of duplex radio communications over said set of channels, including at least one data transceiver, and a set of sensors coupled to each transceiver in said set of transceivers, comprising the steps of:

a) determining when there is unused air time to switch said data transceiver onto the channel to which a particular one of the transceivers in said set of transceivers is tuned;

b) determining when to turn said given data transceiver off based on sensing a demand for the channel to which the given data transceiver is tuned by said particular transceiver;

c) identifying time slots that are unused by said AMPS on each of said channels; and

d) assigning selected unused time slots identified in step (c) for data transmission purposes.

It has to be noted that the AMPS system has been in operation since 1978, and no one suggested the overlay possibility until this patent was filed. Therefore, what appears to be a straightforward evolution of the AMPS system may not have happened if not for the above "invention."

1.2.1 A Brief History

In the early part of 1990, under a project called CelluPlan I, researchers at IBM (Boca Raton, Florida) started looking into employing cellular phones as a means to transfer data (circuit-switched cellular) using a data modem built by Communicate and a cellular phone from Novatel. They soon discovered that modem attributes, terrain, and the noisy airlink affected the reliability of the data transfer. While analyzing the air channel in AMPS cell sites to characterize its behavior, they found that even in the busiest cell sites a significant portion of the time individual RF channels were not used, giving rise to the idea that it may be possible to use the idle time on an RF-channel to transfer data with the ability to hop to different "free" RF channels when voice system began using the channel. IBM and Novatel built a simple prototype to demonstrate the feasibility of the idea and CelluPlan-II was born. Attracting the consent of McCaw to commit to wireless data as an important technology, IBM in partnership with McCaw and subcontracting to PCSI, started specifying the Celluplan-II technology. The task of convincing the carriers to adopt to the IBM-McCaw plan was left to Machaley of McCaw and Moore of IBM, and this effort culminated in the consortium being formed with all major carriers with the exception of Bell South, which had a vested interest in RAM Data. Concerns about possible RF interference of the new technology with the operational AMPS system were dismissed after some internal tests were conducted by McCaw. During 1992, IBM's involvement in the project diminished. A field trial of the initial technology derived from CelluPlan-II was held in the San Francisco Bay Area in the second half of 1992 and early part of 1993. Meanwhile, an improved version of the CDPD technology, which was more tailored to the Internet world, was developed by a team led by Mark Taylor (McCaw) for the CDPD forum. The consortium, later including Bell South, announced the availability of the CDPD specifications in San Jose (1993) in the presence of some major companies such as Sears, which expressed commitment to CDPD as a user. By patenting the technology, IBM ensured that the technology transfer was accomplished at no cost to the carriers and was able to serve as a neutral third party to bring the carriers together to build the infrastructure necessary to support CDPD. The technology specification that defined the airlink as Gaussian minimum shift keying (GMSK) and a data rate of 19.2 Kbps was based on the premise that an inexpensive CDPD modem is key to the success of CDPD. These requirements allow off-the-shelf components available in the cellular industry to be used, minimizing the modem cost.

1.2.2 Overlay on Analog Cellular

Utilizing the unused bandwidth in the AMPS system and the ability to share cell-site hardware are the two factors that gave birth to the CDPD technology as a low-cost alternative to existing private packet data technologies.

In sharing the same set of frequencies allocated to the AMPS, no additional functionality is imposed on the AMPS. The implication is that the AMPS system will continue to allocate and use frequencies assuming that there are no other potential users of its frequencies. It is the responsibility of the CDPD system to continuously monitor ("sniff") the AMPS frequencies and employ the unused frequencies. However, when the AMPS radio is independently assigned the frequency the CDPD system is using, the sniffing subsystem in the CDPD has the responsibility to switch the carrier off within 40 ms, thereby not causing interference with AMPS operation. The CDPD system then has to find another unused frequency. The mobile-end systems (M-ESs) that were tuned to the older frequency will recognize the disappearance of the CDPD carrier and have to identify a "new" available CDPD frequency and continue the session with minimal interruption to the data flow. The CDPD network will assist the M-ESs in providing a variety of information related to CDPD frequency pools and so forth so that M-ESs can acquire CDPD channels efficiently.

The extent to which CDPD is overlaid on the AMPS system ends with the air segment. In the area of sharing the cell-site hardware, one of the most expensive components in a cell site is the real estate for the cell site and the associated antenna subsystem. The CDPD can fully share the AMPS antenna subsystem by suitably combining the RF transmit signals and splitting the RF receive signals. The sharing of power amplifiers (PAs) is also possible depending on the availability and compatibility of the AMPS PAs. In the mobile telephone switching office (MTSO), the AMPS voice calls carried by the T1 links from the cell sites are multiplexed and connected to the public switched telephone network (PSTN). The CDPD network traffic from the cell site is also carried by the same shared T1 links, and in the MTSO this multiplexed data traffic uses a configured router as the gateway to the terrestrial IP/OSI data networks. Figure 1.3 shows the relationship of the CDPD overlay with the AMPS infrastructure.

1.2.3 Dedicated and Shared RF Channels

The frequency allocation scheme in a carrier's AMPS network and the carrier's preference in phasing in of the CDPD network will largely determine the configuration of frequencies for use by the CDPD network. In an AMPS system, each cell will have multiple radios and a set of frequencies assigned to the cell. Similarly, the CDPD system will have a frequency pool configured for each of the cell. In a typical AMPS/CDPD setup, the cell configuration (sectored, omni, etc.), and the RF foot print for the AMPS and CDPD will be as closely matched as possible.

The CDPD use of the frequencies allocated for AMPS fall into two different groups: dedicated frequencies and shared frequencies.

Dedicated Frequencies

These are frequencies assigned by the carrier for sole use by the CDPD network. The CDPD hardware/software do not implement special procedures to monitor the use of these "dedicated" frequencies by other network(s).

In this case, the cooperating AMPS system RF-engineering personnel must manually enforce that the AMPS frequency pool does not include any dedicated frequencies marked for use by the CDPD system. Failure to continually establish this mutually exclusive frequency allocation condition every time a new frequency plan

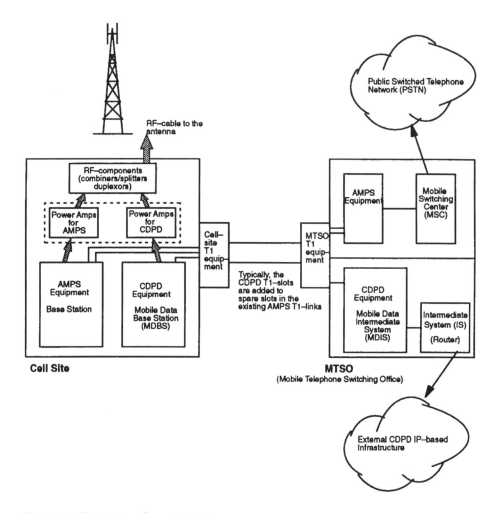

FIGURE 1.3 CDPD overlay with AMPS.

for AMPS is established and implemented will result in the CDPD system interfering with voice.

Shared Frequencies

The frequencies that the AMPS and the CDPD can share are called shared frequencies. The AMPS system is not expected to be aware of any other potential users of its frequencies. It is the responsibility of the CDPD system to use the shared frequencies when the AMPS system is not using it (a call is not active on that frequency) and to relinquish the frequency when the AMPS starts using it before causing any undue interference with the voice system. Special hardware that sniffs voice frequencies and triggers a carrier switchoff within 40 ms from the time of detection needs to be employed in the CDPD network for harmonious sharing of frequencies. This stringent switchoff time requirement eliminates any audible interference in the voice system by removing the CDPD carrier from the air during the initial setup time of the voice call, before the beginning of any voice activity.

CDPD design allows the switched off carrier (on detection of voice activity) to be replaced with another "available" frequency by the base station, and has procedures implemented in the M-ESs to search and tune to a new frequency, making the transition transparent to the application. In CDPD implementations we can find both configurations. In markets where there are frequencies available for exclusive use by the CDPD, the carriers tend to prefer a dedicated channel CDPD setup. This simplifies the CDPD hardware, and hence the cost, as sniffing subsystem hardware is not required. In addition, this simplifies the operational software in the CDPD system and minimizes the chance for interference with the more important (revenue-wise) voice system. The inefficiency in the data transfer mechanism arising from interruptions due to channel "hopping" that occurs in a shared frequency CDPD configuration is also avoided.

1.2.4 Network Connectivity

One of the main objectives in the CDPD network specification and design is to use the functionally proven and extensively deployed existing data network infrastructure as an integral part of the CDPD network.

The open system interconnection (OSI) protocol suite is based on the OSI reference model defined by the International Organization for Standardization and International Electrotechnical Committee (ISO/IEC). Although OSI is predicted to become the primary networking solution of choice in the next several years, existing OSI implementations are few and skeletal. In contrast, TCP/IP-based networks have proven to be operationally successful, and in the last two years have attracted a mass consumer market with Internet access and the World Wide Web possibilities. Thus, the CDPD specification provides facilities to use either the OSI protocol suite with ISO 8473 (connectionless network protocol (CLNP)), or the TCP/IP protocol suite with IP (Internet protocol) as the network layer protocol.

Current M-ES implementations and associated applications exclusively use the TCP/IP protocols. However, communication between CDPD components, primarily mobile data-intermediate systems (MD-ISs), the specification requires the CLNP protocol be used.

1.2.5 CDPD System Performance

The primary component that governs the performance of the CDPD network is the airlink. With the given bandwidth of 30 kHz, and the choice of Gaussian minimum shift keying (GMSK) modulation, CDPD networks operate at a 19.2-Kbps data rate. This data rate has been demonstrated to be superior to those obtained from the current working speeds of existing private wireless networks like ARDIS and RAM Data. However, the private packet service providers are busy upgrading the data rates supported by their networks, and a 19.2-Kbps RAM Data network is being deployed.

The raw data rate translates to the following effective data rates after accounting for the air-link MAC level overhead:

- Link level throughput in a forward direction: 12.7 Kbps;
- Link level throughput in a reverse direction: 11.8 Kbps (based on a half-duplex M-ES, averaged over a single transmission);
- Half-duplex M-ES can send only a limited amount of information in each of its reverse transmissions (274 bits, which is equivalent to approximately 34 octets).

These are approximate throughput values to provide an indication of the overall performance of the CDPD network and apply to one channel stream. The following list indicates why the users have to be careful in interpreting the above throughput values:

- The throughput applies to a CDPD channel stream (one frequency pair). If multiple mobile stations are using the same channel, the mobile units have to share the bandwidth.
- If the same channel stream is shared by multiple M-ESs (all half-duplex), then due to the multi-access protocol constraints, the reverse direction throughput will be less than 11.8 Kbps. The more the number of M-ESs, the less the value of the aggregate throughput of the channel stream.
- The network can be configured to allow the full-duplex M-ESs to send longer streams of information in each of their reverse bursts, thereby achieving some improvement in the reverse channel throughput.
- If multiple channel streams are available in a cell, then the CDPD network can assign M-ESs to different channel streams based on the load-balancing algorithm being implemented. The throughput, in the M-ES's perspective, will then be dependent on total channel utilization and the effectiveness of the load balancing.

Further, the above throughput applies to the link-level frames. Accounting for the link level and network-level overhead would leave the effective application throughput 10% to 20% less than the figures shown above. Other aspects related to throughput to take note of are as follows:

- CDPD operation allows for header compression when using the TCP/IP or the OSI protocols. A standard 40-octet header in each packet in the TCP/IP session is compressed to an average of 3 octets. Throughput as seen by the application will have marked improvements if header compression is enabled.

- CDPD Specification 1.1 specifies V42.bis compression by the network layers. Applications that require transfer of large volumes of data and do not employ compression in any form at the application level can again experience larger effective throughput. The CDPD networks currently in operation use only one channel stream per cell. Therefore, the effective distribution of M-ES loads across multiple channel streams has not become a hotly debated topic yet.

1.3 CDPD APPLICATIONS

Cost and available bandwidth are two key factors that will have an impact on the type of applications that will use the CDPD technology for wireless data transfer. Two major categories of applications can be identified that can be expected to dominate the CDPD market: embedded systems with bursty data transfer requirements and handheld interactive computing [4–8].

1.3.1 Embedded Systems With Bursty Data Transfer Requirements

CDPD technology is ideally suited to applications where small amounts of data need to be transferred from remote locations to a central system. These applications benefit from the fact that the applications can be "registered" with the CDPD network once during the initial power-up, and can send data without any delay when necessary. The continuous connectivity maintainable by the application without incurring any cost makes CDPD different from circuit-switched cellular where maintaining a continuous connection will be expensive; further, for bursty data transfers, the circuit connection has to be set up and cleared, adding delay in transferring data.

Application types that will fall into this category include both manned and unmanned types. Examples are as follows:

- *Telemetry.* Monitoring of pipelines and so forth as part of interconnection of supervisory control and programmable logic controllers (PLCs) across CDPD networks.
- *Vehicle tracking.* As an inexpensive alternative implementation for satellite-based global positioning system (GPS), vehicles/trucks can use CDPD-based position location system.
- *Credit card authorization.* CDPD can replace the traditional wireline connections used in the credit card authorization system used in the stores. The CDPD system will be faster (does not have to set up the call) and allow authorization stations to be set up anywhere (provided there is CDPD coverage).

1.3.2 Handheld Interactive Computing

This group of applications represents a traditional computing setup where a user and handheld computing equipment are involved. These full-function mobile personal computers are likely to have full screens, keyboards, and different disk configurations (hard and removable and PC-card features). Tablet computers, which will replace multipart forms and possess signature capture capability, will dominate some sectors of the field application market. The following major application categories can be identified within this group.

Personal Messaging

With corporate users increasingly using remote access to enterprise networks to share information and to coordinate/manage activities via electronic mail, peer-to-peer wireless messaging will soon become a business standard. The average consumer who is currently enjoying the convenience of cellular voice and pagers is soon going to discover the similar benefits of electronic messaging and instant access to information (stock quotes, hotel information, etc.), creating a surge in demand for easy-to-use new wireless products and the infrastructure capacity to support the data transfer requirements.

Field Automation

Applications where productivity is increased and the service quality is enhanced by the ability to instantly query or modify information remotely from the field will increasingly employ wireless technology. Police officers, real estate/insurance agents, sales representatives, and field service technicians are examples of the kinds of workers who work away from their offices, are highly mobile, and are bound to benefit from the capability to access information instantly. Police departments in some areas (Connecticut and Pittsburgh) have already adopted or are experimenting with CDPD already.

Computer-aided dispatching (CAD) is another market that could benefit from CDPD. Continuous communication is possible between the office worker and the dispatched field personnel. Taxicab firms and the trucking industry are increasingly embracing CAD, and CDPD can be the cost-effective technology for the needs of these industries.

Internet Access

The popularity of the Internet among computer users has risen sharply in the last year. Accompanying this popularity is the availability of instant information in fields ranging from newspapers to music. The potential of applications and markets built around World Wide Web (WWW) is enormous. At best, a mobile user will be able to

quency spectrum allocation, frequency reuse, and co-channel interference aspects of AMPS have a direct impact on the CDPD overlay network.

2.1.1.1 Frequency Spectrum and Channel Spacing

A 40-MHz block of spectrum, with two 20-MHz blocks for forward and reverse channels separated by 45 MHz, were initially allocated for AMPS. These bands were split into 666 30-kHz wide channels, and a pair of channels, one from each band, represent an RF channel for AMPS operation. An additional 10 MHz of noncontiguous block of spectrum providing 166 channels were later added to the original AMPS RF channel set as shown in Figure 2.1.

The mobile station (handset) transmits on the lower frequency band (reverse channel) and the base station transmits on the higher frequency band (forward channel). The RF channels are grouped into A and B blocks for assignment to two different carriers, a wireline carrier and a wireless carrier, for a geographic area as mandated by the FCC ruling that came into effect in 1981.

2.1.1.2 Frequency Reuse and SATs

Frequency reuse refers to the use of same radio channels in different geographic areas separated by sufficient distances so that the interference between them is not disrup-

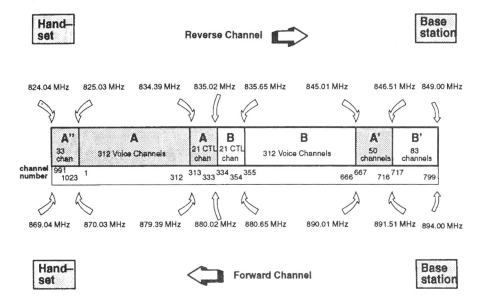

FIGURE 2.1 AMPS frequency spectrum and channels.

tive to the operation in any one site. The cellular concept then becomes fundamental to frequency reuse. Instead of covering a large local area with high-power transmitters at a high elevation, the service provider can employ multiple transmitters with moderate power, each covering a subarea or a "cell." Given a limited number of voice channels, it becomes clear that the size of cells and the frequency reuse strategy based on cell clusters define the overall traffic capacity of the network (i.e., in one coverage area the cellular system can handle a number of simultaneous calls that exceeds the number of allocated frequencies).

For omnidirectional transmitting antennas, the cell boundaries, which are defined by contours of constant signal strength representing the minimum received signal strength indication (RSSI) required to make a successful call, will be roughly circular. However, a hexagonal approximation is typically assumed to simplify theoretical and analytical radio propagation studies.

Figure 2.2 shows the different frequency reuse patterns when the cell clusters that have unique frequency sets consist of 4, 7, 12, and 19 cells, respectively. If D is the distance between the center of two co-channel cells, R is the cell radius, and K is the number of cells in cluster, it can be shown that

$$D = \sqrt{(3K)} \cdot R$$

For the different cluster configurations shown, the corresponding values are as follows:

$$
\begin{aligned}
D &= 3.46R & K &= 4 \\
&= 4.60R & K &= 7 \\
&= 6.00R & k &= 12 \\
&= 7.55R & k &= 19
\end{aligned}
$$

Theoretically, a large K that provides a large co-channel cell distance is desired. This, however, will result in low number of channels per cell. But the co-channel interference calculations reveal that a D/R ratio is required to obtain a carrier to noise ratio (C/I) of 18 dB in an omni arrangement [1]. This indicates that a 7-cell reuse pattern with a D/R ratio of 4.6 is the pattern with the lowest number of cells/cluster that satisfies the co-channel interference requirement.

The supervisory audio tone (SAT) ensures the maintenance of a reliable transmission session between the mobile station and the selected ground station. The SAT is one of three frequencies: 5,970, 6,000, and 6,030 Hz. Different SAT tones are assigned to cell clusters in close proximity, so that co-channel interference can be identified by the mobile and the base station. During call establishment and handoff, the SAT for the particular session (groundstation, RF channel) is established. If the received signal at the mobile station or at the base station has an incorrect SAT code,

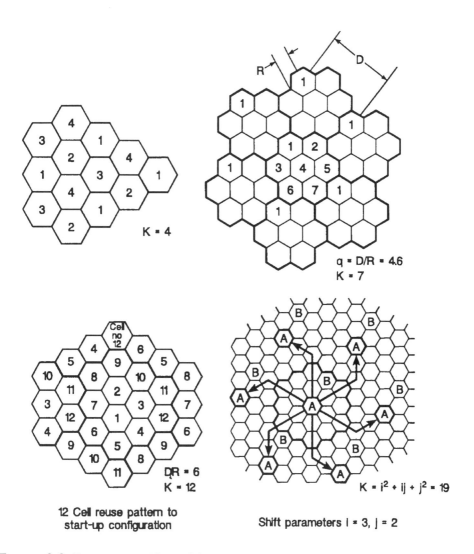

FIGURE 2.2 Frequency reuse. (*Source:* [1].)

then it is assumed to be co-channel and the audio is muted so that voice from another session is not overheard.

2.1.1.3 *AMPS Signaling*

The 21 control channels in each block of the spectrum are used by the base station and the mobile station to exchange information relating to the setting up of calls. Once the call is in progress, data signaling is used on the traffic channel and the speech path is muted during these bursts of data to prevent interference with speech, and 10-Kbps FSK signaling is used in the control channels and during messaging in the traffic channels. Control channel messages include the following:

- Overhead messages, containing general data on the local system for all mobiles to receive;
- Mobile station control messages that are sent specifically addressed to a mobile, which includes power level, voice channel information, SAT frequency, and so forth;
- Control filler messages to ensure a continuous stream of data in the forward control channel.

All data are encoded into Bose-Chaudhuri-Hocquenghem (BCH) code words, a well-known class of multiple error correcting cyclic codes. A (48,36) BCH code is used in the control channel, and a (40,28) code is used in the voice traffic channels (blank and burst FSK digital messages), both involving 12 parity bits. In addition, data words are repeated several times (11 in the forward voice channel and 5 in the reverse voice channel) to ensure that commands and responses are not missed.

The mobile station also uses a signaling tone (ST) of 8 kHz for additional control using the voice channel. The ST is used exclusively by the mobile station for the following:

- To terminate a call (1.8 seconds of tone);
- To request to send dialed digits (400 ms of tone);
- As a confirmation of handoff request (50 ms of tone);
- As an alert (continuous tone, removal of tone under offhook).

Other signals sent on the voice channel include supervisory audio tones (SATs) and dual tone multifrequency (DTMF) signals.

A modified version of the AMPS system, called narrowband AMPS (NAMPS), which subdivides AMPS voice channels into 10-kHz blocks to obtain a three-fold capacity increase, has also been approved by TIA (IS-88, 89, 90) for deployment. The NAMPS system continues to use the same control channels (30-kHz wide) defined in the AMPS system.

2.1.1.4 Large-Capacity Cellular Systems

The different types of wireless analog cellular telephone systems installed across different parts of the world are generally not compatible and differ in the aspects of bandwidth of RF channels and the frequency spectrum of the channels used. All of the systems listed in Table 2.1 use frequency modulation (FM) as the modulation technology for the airlink.

TABLE 2.1

Comparison of Wireless Analog Telephone Systems

Parameter	Japan	U.S.	U.K.	Scandinavia	West Germany
System	NTT	AMPS	TACS*	NMT†	C450
Channel width (kHz)	25	30	25	25	20
# of channels	600	666 + 166	1,000	180	222
Coverage radius (km)	5–10	2–20	2–20	1.8–40	5–30
Voice:					
Modulation	FM	FM	FM	FM	FM
Frequency Deviation (kHz)	±5	±12	±9.5	±5	±4
Control:					
Modulation	FSK	FSK	FSK	FSK	FSK
Frequency Deviation (kHz)	±4.5	±8	±6.4	±3.5	±2.5
Data rate (Kbps)	0.3	10	8	1.2	5.28

* TACS: Total access communication systems.
† NMT: Nordic mobile telephone system.

Clearly, in many parts of the world, the CDPD system as applied to the AMPS cannot be used with the non-AMPS technologies. However, the CDPD technology can be easily adapted to operate as an overlay in the non-AMPS systems mentioned above, which is similar to AMPS but with different channel widths and operating in different frequency bands.

2.1.2 Digital Cellular Systems

Digital cellular systems evolved from the first-generation analog cellular systems. The second-generation digital cellular systems are based on a time division multiple access (TDMA) scheme. Several standards, such as IS-136, GSM, and Japanese digital cellular (JDC), exist that differ in modulation scheme, error-correction code procedures. Using digital technology, these systems aim to increase the circuits (calls) per carrier without degrading the voice quality of the analog cellular system. Third-generation digital systems are based on such technologies as code division multiple access (CDMA) and enhanced-TDMA (E-TDMA), and they aim to further increase the call-carrying capacity from available frequency spectrum.

2.1.2.1 TDMA (IS-136) North American Digital Cellular

Digital cellular technology based on a combination of FDMA and TDMA technology, in addition to providing greater capacity (three or six times that of AMPS), promises enhanced authentication, privacy, low-power handsets, more consistent voice quality, and fewer dropped calls compared to the AMPS system. With the increasing popularity of CDMA, it is not clear whether the IS-136 will establish itself as the carrier's choice for deployment.

The first generation dual-mode (AMPS/IS-54B) systems use the existing control channels; that is, previously reserved fields are used for digital channel assignment. A later revision, IS-136, includes digital control channels and is backward-compatible with IS-54B and AMPS. The digital control channel uses the same modulation and format as the digital traffic channel. Yet another specification, PN-3388, extends the IS-136 by specifying the operational requirements for providing seamless cellular service between 800-MHz and 1,900-MHz frequency bands.

The RF channel spacing is the same as in AMPS, which is 30 kHz. The carriers typically choose selected AMPS channels to be operated in the digital mode as they phase in the digital technology so that the users who continue to use analog-only phones are not denied service. Each digital channel operates at 48.6 Kbps and uses a simple frame structure shown in Figure 2.3.

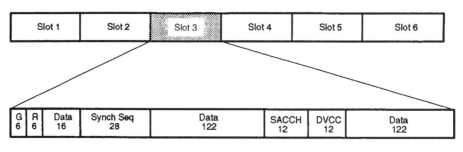

Mobile to Base (Reverse)

Base to Mobile (Forward)

G: Guard Time
R: Ramp Time
DVCC: Digital Verification Color Code
RSVD: Reserved for Future Use

FIGURE 2.3 IS-54 time slots and frames. *(Source:* [2].)

The 40-ms frame is divided into six 6.67-ms slots, each time slot carrying 324 bits, which includes the following:

- 260 bits of user data;
- 12 bits of system control information—slow associated control channel (SACCH);
- 28 bits of synchronization signal;
- 12 bits of digital verification color code (DVCC);
- In the mobile to base direction (reverse), 6 bits guard time when no energy is transmitted followed by a 6-bit ramp interval for the transmitter to reach its full power.

The system specifies six different synchronization patterns, one for each time slot, allowing the receiver to select the assigned time slot. The digital color code provides the same functionality as the supervisory audio tone (SAT) in the AMPS system that allows the receiver to distinguish adjacent channels that result from the frequency reuse structure.

Full-rate channels operate at 13 Kbps and occupy two slots in a 40-ms frame. Speech coder is a vector sum excited linear prediction (VSELP) type. Half-rate channels, which will employ 4-Kbps vocoders, will need only one slot per frame and will operate at 6.5 Kbps, providing six times the capacity of an equivalent AMPS system. IS-136 uses a linear modulation technique, differential quartenary phase shift keying (DQPSK), which is bandwidth-efficient but is not a constant envelope modulation technique. The specific method of modulation is $\pi/4$ shifted DQPSK with root cosine rolloff filtering at the transmitter and receiver, the rolloff factor of 0.35 being obtained by placing a zero at 16.4 kHz in the baseband spectrum. The bandwidth efficiency is 48.6 Kbps/30 kHz = 1.62 b/s/Hz, providing a 20% improvement over global system for mobile communications (GSM), a widely used system in Europe.

The VSELP coding is known to distort music and also is problematic for dual-tone multifrequency (DTMF) response systems like voice mail, fax, and remote control answering machines. DTMF tone generators can be installed at the MTSO to circumvent this problem. Vocoder distortion also obstructs the use of modems over the digital voice channels.

Because the IS-54B does not define data services, no standards exist to provide data service over digital cellular. However, infrastructure equipment manufacturers have implemented proprietary protocols to transmit fax and data over digital cellular systems. In IS-136, provision for digital asynchronous data transfer and Group 3 fax are included, defined by the new standards IS-130 and IS-135. Fast, error-free file transfers at 9,600 bits per second (uncompressed) are possible with these new TDMA standards. Further, a short message service (SMS) feature defined as part of IS-136 allows the IS-136 handset to function as a pager. In addition, mobile-originated or mobile-terminated short messages can be exchanged with acknowledgments using the defined fields in the digital traffic channel as well as in the digital control channel.

2.1.2.2 E-TDMA

E-TDMA, a Hughes Network Systems (Germantown, MD) developed technology, is an extension to the IS-54B standard and uses the same protocol architecture, frame formats, and control mechanisms. In IS-54B, each call is assigned a fixed time slot in a specific channel, which remains unchanged for the duration of the call. In contrast, E-TDMA uses dynamic assignments of the call-to-time slot mapping while the call is in progress. Further the time slot switch need not be restricted to be in the same RF channel but can be selected from a configured pool. By identifying the periods of silence in each conversation, E-TDMA reduces the amount of speech data that needs to be transported. This digital speech interpolation (DSI) technique is illustrated in Figure 2.4.

Voice activity detection (VAD) capability is crucial for the implementation of the E-TDMA technology. For IS-54B and E-TDMA operation, time slots have to be made available to carry control channels. Figure 2.5 further illustrates different scenarios when a larger number of digital channels are made available for E-TDMA operation. Examples of the number of control slots required and the possible number of voice calls that can be supported using the remaining time slots for a DSI pool containing 3, 8, 12, and 19 radio channels are shown in this figure. These represent gains of approximately 7, 9.5, 10.3, and 11 for the different pool sizes over the AMPS capacity.

With this technology, the same voice channel (30 kHz) can support a mix of half-rate IS-54B calls and full-rate IS-54B calls and still have slots allocated for use in the E-TDMA pool. E-TDMA also has a "soft" capacity limit; the maximum number of simultaneous users is not fixed. However, as the number of users increases, so does the risk of losing occasional voice spurts.

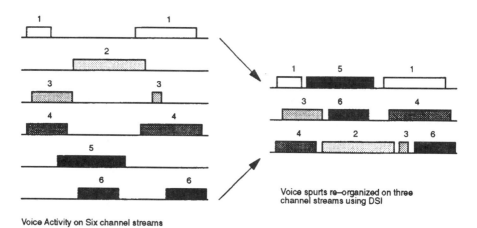

Voice Activity on Six channel streams

Voice spurts re-organized on three channel streams using DSI

FIGURE 2.4 Digital speech interpolation for E-TDMA. (*Source:* [3].)

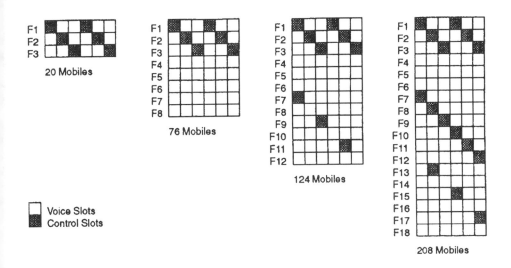

FIGURE 2.5 DSI pools with increasing numbers of digital channels. (*Source:* [3].)

2.1.2.3 Spread Spectrum Communications

The spread spectrum modulation uses a transmission bandwidth far greater than information bandwidth or the data rate of a user. Spectrum spreading can be achieved by one of two techniques: direct sequence spread spectrum and frequency hopping spread spectrum.

Direct Sequence Spread Spectrum (DS-SS)

In DS-SS, each data bit is symbolized by a large number of coded bits independent of the data stream, called *chips*. The collection of coded bits is also called the spreading sequence, which possesses the pseudo-randomness property that allows multiple sources with different spreading sequences sharing the same bandwidth.

Frequency Hopping Spread Spectrum (FH-SS)

In FH-SS, the wider spectrum will be split into several hopping channels (also called chips), and a hopping pattern defined by a pseudo-noise (PN) sequence will be used. If two or more hops are performed per symbol, the technique is called fast FH, and if two or more symbols are required before a hop, it is slow hopping.

The multiple access technique used in cellular systems is based on the direct sequence type and is popularly known as code division multiple access (CDMA).

Narrowband CDMA (N-CDMA) Developed by Qualcomm, the N-CDMA technology has been adopted as IS-95 digital wireless standards by the Telecommunications Industry Association (TIA).

CDMA proponents point to the following weaknesses in the traditional techniques:

- Transmission of information is not continuous in every channel.
- Cellular technology and its frequency reuse strategies reduce the number of frequency channels per cell by the reuse factor (cluster size).
- Presence of multipath fading resulting from phase cancellation between different propagation path.

The modulation and data transformation procedures carried out in the forward and reverse channel in a typical CDMA implementation are shown in Figure 2.6.

Each CDMA channel is 1.8 MHz wide with a 1.25-MHz voice channel surrounded by two 275-kHz guard bands. For proper reception of the CDMA signals at the base station, the received signal strengths must be within 1 dB of each other. This is called the near-far problem, and accurate closed loop power control is necessary. However, this requirement has a positive side effect in that it allows the mobile to transmit at low power levels.

N-CDMA offers soft handoff, graceful capacity degradation, the use of rake receivers to deal with multipath propagation, and an inherent form of digital speech interpolation. Soft handoff exists when a call temporarily exists on two adjacent base stations, made possible by the use of the same frequencies and different code sets. By choosing the best signal from the adjacent cells at each instant, the system minimizes the risk of dropping calls due to premature or late handoffs. In CDMA, an additional user in the channel amounts to an increase in the background noise for the other users of the channel. When the noise level increases to a level that is unacceptable for a user (at a high channel utilization) the user will decide to terminate the call, introducing an element of self-regulation. The DSI is used to alter the transmission rates, thereby eliminating the risks of losing voice spurts. However, the minimum transmission rate is 1 Kbps, set by the power control requirement.

The strategy being adopted for implementation of CDMA is to make it the final stage in the upgrading of wireless technology. For example US West is intending to divide the spectrum into 12.5 MHz for AMPS, 10 MHz for CDMA, and 2.5 MHz for N-AMPS.

Broadband CDMA (B-CDMA) A higher performance version of Qualcomm's technology has been proposed by InterDigital as a solution for digital cellular. It is designed to coexist with existing cellular, thereby allowing the carriers to deploy the system without sacrificing the capacity of the existing voice. Adaptive notch filters are used to attenuate narrowband users, allowing the coexistence.

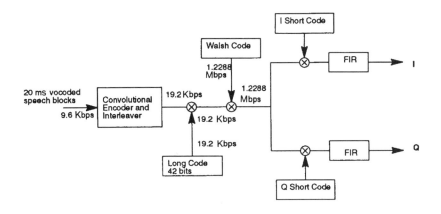

Forward Channel: QPSK modulation

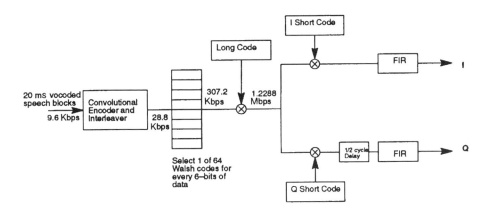

Reverse Channel: O–QPSK modulation

FIGURE 2.6 CDMA physical level procedures.

2.2 CIRCUIT-SWITCHED CELLULAR

With about 20 million subscribers using the cellular telephones and the extensive infrastructure already in place, the analog cellular medium appears to be the best platform for providing mobile data. Similar to using two landline modems to establish a connection, cellular modems can be used to transmit and receive data on the voice channel of the cellular medium. Widespread use of this medium for data exchange is inhibited by the following drawbacks inherent to the technology:

- Most cellular phones do not have interfaces to connect auxiliary equipment such as modems or fax machines. Even the ones that do, the modems and adapters are not compatible. For interoperability, the cellular modem or adapter must interface to the bus of the cellular telephone. Although some companies such as NEC and Mitsubishi publish open specifications that are supported by cellular modem and adapter vendors, many cellular phone manufacturers have closed architectures.
- AMPS voice channels have smaller bandwidth than their landline counterparts allowing data transfer only up to 4,800 bps. Further, the loss of carrier experienced during the cell handoffs of 1–2 seconds will ordinarily cause the modem to hang up. This can be overcome by setting the modem's internal timer to 5 seconds or more.
- Cellular voice channels are susceptible to channel impairments inherent to mobile radio such as co-channel interference, adjacent channel interference, and multipath fading. These impairments manifest during initial call setup by disrupting the handshake and causing the modem to abort the call or to drop to a lower baud rate; during data transfer channel impairments can cause loss or corruption of data.

To lessen the impact of the above drawbacks several special airlink protocols for use in the circuit-switched cellular environment have been developed. Some of the well-known protocols and a brief description of their characteristics follow.

- *Microcom networking protocol (MNP).* The MNP-10 protocol incorporates most of the features that are contained in the lower level MNP modems such as error checking and retransmission, adaptive packet sizing, and data compression. The error control protocol contains a feature similar to selective reject (SREJ) in link access protocol-balanced (LAPB) mode, where only the erroneous packet needs to be retransmitted. The Class-10 contains several adverse channel enhancements designed specifically for cellular and negotiated speed upshift. Unlike V.32 class of modems, which try at the highest speeds first and move towards lower speeds until a reliable connection is achieved, the MNP-10 starts at 1,200 bps (connect at a lower rate) and increases the rate, making modems connect faster. The robust auto-reliable mode allows multiple attempts at negotiating error-controlled connections. The adaptive packet assembly option starts out with short packets and lengthens the packet size while the error rates are low. The dynamic transmit level adjustment modifies the transmit signal strength to match the cellular link.

- *Enhanced throughput cellular (ETC).* Built as an enhancement to the V.42 standards, AT&T Paradyne's ETC protocol-based modem contains similar features offered by MNP-10, but has been shown to obtain higher throughput than the MNP-10. One main advantage with this modem is that it can interwork with any V.32bis/V.42 modem and still get "most" of the throughput benefits.

- *ZyCellular.* Running a proprietary protocol, the cellular modems manufactured by ZyXEL have been shown to sustain data rates of up to 14.4 Kbps. The proprietary nature of the protocol necessitates the use of a compatible modem at the other end of the link. Further, the modem requires an ac supply and an adapter, making its portability questionable.

- *Cellular data link control (CDLC).* Developed and mainly used in the United Kingdom, CDLC is a full-duplex layer 2 protocol modeled on the HDLC standards. CDLC employs techniques including Reed-Solomon forward error correction (FEC), BCH codes, and bit interleaving to obtain increased performance over noisy cellular channels.

The cellular service providers have taken steps to ease some of the problems faced by the circuit-switched cellular users. The architecture shown in Figure 2.7 is used so that the mobile user's remote end need not have to be equipped with a compatible modem to the one used at the cellular phone.

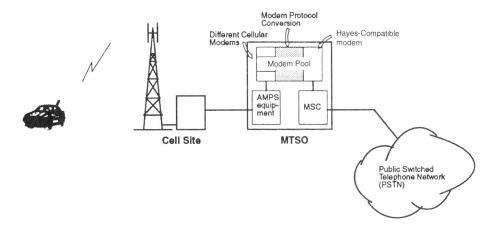

FIGURE 2.7 Organization of modem pools in the MTSOs. (*After:* [4].)

The user is provided with a prefix to the conventional number that the user wishes to access to obtain the traditional wireline data service. When the prefix is dialed, the cellular modem is connected to its compatible modem in the carrier's modem pool located in the mobile telephone switching office (MTSO). This cellular modem pool is connected back to back to a Hayes-compatible modem, which then uses the carrier's public switched telephone network (PSTN) to establish the intended connection to the remote data service that contains a Hayes-compatible modem as its front end.

Xmodem or Kermit, which use a 128-byte packet size, are two of the file transfer protocols better suited for cellular applications. The more popular Ymodem or Zmodem protocols with their 1-KB packet size have been found to perform worse because of the inefficient error recovery of the protocol with larger packet sizes.

Another major disadvantage in circuit-switched cellular is the cost. The charge is based on connection time. Therefore, for short bursty transactions, the connection setup time will be a significant fraction of the total transaction time and will not prove to be cost effective. In radio packet networks, the cost will be based on the amount of data and is therefore likely to be less.

2.3 WIRELESS DATA IN GSM-BASED SYSTEMS

The global system for mobile communications (GSM) is a protocol standard defined by CEPT for use in digital land mobile radio networks and is widely used in Europe.

Similar to IS-54, GSM is based on a cellular concept, but uses the frequency bands 890–915 MHz for reverse channels and 935–960 MHz for the forward channels. The RF band is split into 124 pairs of channels each 120-kHz wide. GSM, like IS-54, uses a burst mode transmission in the reverse direction (mobile to base). The GSM air interface access technique is a combination of frequency division (200-kHz carrier spacing) and time division with eight logical channels per carrier (full rate). During a call, each terminal has access to a two-way digital traffic channel and a separate two-way control channel. Figure 2.8 shows the organization of these channels.

A GSM multiframe of 120-ms duration is divided into 26 frames, each containing eight time slots. In the multiframe, 24 frames carry user information traffic in logical traffic channels and the other two frames contain control channels.

Two bursts of 58 bits account for most of the transmissions in a time slot. One flag bit represents whether the contents are speech information or other data. There are 26 bits of equalizer training sequence and the time slot begins and ends with three tail bits, all logical zeros. An 8.25-bit guard time prevents overlapping of bursts from different mobile terminals. The data transmission rate can be seen to be 156.25 bits per 577 μs, which represents 271 Kbps.

GSM uses the same modulation scheme that is used in CDPD, which is Gaussian minimum shift keying (GMSK). However, the Gaussian filter has a 3-dB cutoff

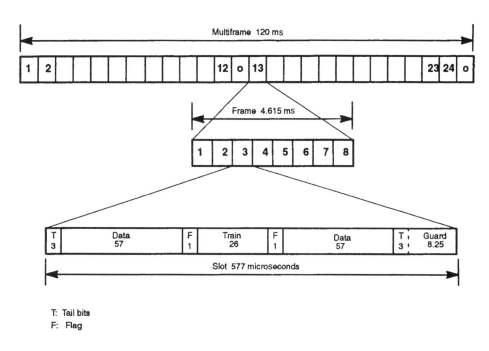

T: Tail bits
F: Flag

FIGURE 2.8 GSM multiframe, frame, and time slot structure. (*Source:* [2].)

frequency of 81.25 kHz, which is 0.3 times the bit rate (a BT product of 0.3) as opposed to CDPD where the equivalent bandwidth, bit rate product is 0.5.

Support for data is designed as an intrinsic part of the GSM protocol. Two basic data transfer modes are available in GSM.

- *T-mode.* Error correction is entirely done by the forward error correction codes defined as part of the radio interface transmission. In T-mode, the path from the terminal, or more precisely the terminal adaptation function (TAF), and the GSM network internetworking function behaves as a synchronous circuit with constant throughput and deterministic delay. Instead of opting for simple rate adaptation schemes to obtain lower rates, the GSM defines three different intermediate rate modes so that increased quality of service can be obtained using the FECs at lower rates.
- *NT-mode.* In this mode, a retransmission scheme is used to recover from "frame" errors. In the NT-mode, the transmission on the GSM connection is considered as a packet data flow. Data is segmented into 200-bit frames and in addition to employing the same forward error correction scheme as in the T-mode, the frame includes redundancy bits for the receiver to detect remaining errors. Total frame length including the FEC and redundancy bits is 240 bits. These redundancy bits form the basis for the radio link protocol (RLP), which

similar to other link level protocols and uses sequence number based retransmission techniques for recovering from FEC-uncorrectable errors.

The available data connection types are summarized in Table 2.2.

TABLE 2.2
The GSM Transmission Connection Types

Connection Type	Quality of Service	Delay (Two-Way)
TCH/F9.6 T	Low	330 ms
TCH/F9.6 NT	High	>330 ms
TCH/F4.8 T	Medium	330 ms
TCH/F2.4 T	Medium	200 ms
TCH/H4.8 T	Low	600 ms
TCH/H4.8 NT	High	>600 ms
TCH/H2.4 T	Medium	600 ms

Source: [5].

2.4 PRIVATE PACKET SYSTEMS

ARDIS and RAM Data fall into the category of private wireless packet data networks. These packet networks, similar to CDPD, have the following advantages compared to circuit-switched cellular:

- Packet networks enable multiple users (mobile-end systems) to share the same channel;
- They provide protocols for detecting and correcting errors using FECs and retransmission;
- The cost of network access is based generally on the amount of information transmitted and not on the connection time; channel errors causing retransmission will not impact the cost incurred by the user.

One of the main objectives in the design of the private packet networks is to provide in-building coverage to handheld devices. Power levels of the early ARDIS and RAM devices were in the 2W range, and the next generation is expected to be 1W or less.

2.4.1 RAM Data

RAM Mobile Data was formed in 1989, and BellSouth became a major partner in 1992. RAM Data claims that the service is available in 6,300 U.S. cities covering 210 metropolitan areas and that its network reaches 90% of the U.S. business population.

The protocol used in the RAM Data network is Mobitex, developed by Ericsson and used widely in Europe. It is an open specification for two-way land mobile communications but implemented as a data only service in the United States. As an adaptation of an SMR system, RAM Data uses 10 to 30 channels in each of the 100 top MSAs. The channels are located in the 935–940 MHz (forward) and 896–901 MHz (reverse) bands.

The Mobitex protocol, like the CDPD specifications, uses GMSK as the modulation scheme in the airlink. Mobitex's choice of 8 Kbps and a BT product of 0.3 (as opposed to 0.5 for CDPD) allow tighter channel spacing (12.5 kHz). With greater intersymbol interference, the system has limited tolerance to noise and distortion. The narrower channel also limits Mobitex's tolerance to frequency offsets between units. Several manufactures offer single-chip baseband GMSK modems facilitating the implementation of a GMSK system using standard FM radio topologies.

Mobitex is a hierarchical network consisting of network control center, main exchanges, area exchanges, and base stations. Traffic flows only as high up the hierarchy as necessary; switching can be done locally at any node level. Billing and mobile tracking information, however, is always sent to the network control center.

2.4.2 ARDIS

Advanced Radio Data Information Service (ARDIS) packet data network was built by IBM and Motorola in 1984, mainly to provide in-building coverage. The ARDIS system operates in a portion of the spectrum just below the cellular telephone channels, with each of the channels having a bandwidth of 25 kHz, and uses a proprietary protocol. However, much of the ARDIS system operates on a single channel, which is augmented by additional channels in areas where higher capacity is required.

By using overlapping cells using the same frequency, an increased coverage of buildings is achieved. But the base station transmitters are frequently turned off (for up to a second) to prevent interference, reducing the overall capacity of the network. Two-level FM modulation and a packet size of 256 bytes are used and the radio frequency links run at 4,800 bps. Approximately 60% of the throughput is consumed by the overhead, resulting in an actual application data throughput of about 2 Kbps. The 25-kHz wide duplex channels are separated by 45 MHz. More than 1,300 base stations, each covering an area of 15–20 miles radius, are claimed to reach 90% of the U.S. business activity. The base stations are connected to 1 of 35 radio node controllers, which in turn are connected to a triply redundant message switch based on tandem computers. Each message from a mobile system has to travel all the way up the hierarchy to be processed by the message switch.

Upgrading of ARDIS base stations to run at 19.2 Kbps using Motorola's RD-LAP protocol and four-level FM modulation is planned. This upgrade is intended mainly to increase the capacity of the network and not to increase the throughput of individual users.

Motorola's Envoy (personal communicator) features an integrated wireless modem for the ARDIS packet radio network. With access to a mailbox from the RadioMail, users can exchange electronic mail over ARDIS with private/public mail systems including cc:Mail, AT&T Mail, MCI Mail, America Online, and CompuServe. The Envoy benefits from General Magic's Magic Cap operating system, which provides the platform to build a host of communication and personal management applications.

2.4.3 Summary of Private Packet Networks

A brief comparison of the private packet networks and the circuit switched cellular is given in Table 2.3.

TABLE 2.3
Comparison of Packet Networks

Parameter	Circuit-Switched	ARDIS	RAM
Metro areas	740	409	216
Base stations	13,000	1,300	870
Subscribers	90,000	38,000	12,000
Modem price	$300	$800	$400

Although the number of base stations is not a good measure of the coverage provided, the coverage advantage of the circuit-switched cellular (cellular voice) over the ARDIS and RAM networks is well established. However, the tariff structure, connection reliability, and better in-building coverage have assisted the private packet networks to gain increasing market share.

2.5 SPECIALIZED MOBILE RADIO

Specialized mobile radio (SMR) is a trunked system operating in the 800 MHz to 900 MHz range. It is privately owned and primarily used for data transmission in the blue collar industries such as taxi dispatching and fuel oil delivery management.

Racotek has designed and built a transaction-oriented service called RacoNET that overlays existing SMRs. Trunked channels to over 1,100 locations across the United States are claimed to be in use. The proprietary Mobile Network Operating System (R/MNOS) provides an open architecture that allows portability across different radio infrastructures and host platforms. Radio modems use phase shift keying (PSK) and run at 4,800 bps.

Prior to 1990, SMRs were classified as unregulated private carriers and were barred from constructing cellular networks. Nextel, which has been acquiring SMR licenses nationwide, has petitioned the FCC to allow it to build a digital cellular network and is focusing on building an enhanced SMR (ESMR) network. The ESMR network planned is based on a cellular TDMA technology developed by Motorola called Motorola Integrated Radio System (MIRS). A thirtyfold increase in system capacity is attained by adopting frequency reuse techniques ($\times 5$) and dividing each SMR 25-kHz channel into six time slots. ESMR, in addition to providing circuit-switched and packet-switched mobile data, will also support roaming.

Another competing technology based on SMR is the digital-SMR (DSMR). DSMR is noncellular and divides the 25-kHz SMR channel into three or six voice channels. Yet another alternative DSMR based on a frequency hopping multiple access (FHMA) spread spectrum technology has been developed by Geotek. Thirty times the capacity of the analog SMR is claimed with the use of 40 narrowband channels of 12.5-kHz bandwidth. A single large sectorized cell is used (35-mile radius), resulting in only one eighth of the cost of a comparable TDMA network.

2.6 PERSONAL COMMUNICATION SERVICES

The personal communication services (PCS) is the aggregate of existing and new services that promise near-universal access to mobile telephony, messaging, paging, and transfer of data. The salient features that are associated with PCS are as follows:

- A location-independent personal number that can be used provide telephony, fax, and data services;
- Low-power terminals whose battery life far exceeds those of current handsets;
- Low cost, and one handset for home, office, and on the street;
- High quality of service, and increased security provisions. The features achievable with current allocation of PCS spectrum in the 2-GHz region include wireline voice quality and high data throughput. Future possibilities include wireless multimedia as indicated by WINFORUM's Supernet proposal and Apple's wireless NII proposal.

Cordless phones whose origins are based in United Kingdom are considered as having provided the seed technology that PCS is based on. The PCS concept has, however, grown to encompass wider objectives, as listed above.

2.6.1 Evolution of Cordless Phones—The CT Series

The cordless phone concept has its roots in United Kingdom. The first-generation cordless phones (CT1) used analog techniques and operated in the high frequency (HF) band. These had limited range, were susceptible to noise and interference, and offered little privacy.

The CT2 was defined around FDMA/TDD technology where the FDMA provided distinct channels for conversation and the TDD provided the time slots for talk and listen times during a conversation. For CT2, the carriers, spaced at 100 kHz, each carry one conversation with the time slots allocated as shown in Figure 2.9.

Each slot has 64 bits of user information and 4 bits of control information. An alternate multiplex arrangement allows only 2 bits of control information and a larger guard time between slots. CT2, in addition, has two guard times of 8 bits duration providing a bit rate of 72 Kbps. CT2 uses binary phase shift keying (BPSK) and has a bandwidth efficiency of 0.72 b/s/Hz.

CT2 was restricted to one-way calling and did not support handoffs. These technical limitations and the resulting failure of the CT2 services gave rise to two enhanced versions: CT2Plus (Northern Telecom) and CT3 (Ericsson).

CT3 uses 1-MHz–wide carriers divided into 16 time slots. The CT3 handset temporarily sets up two parallel voice paths when a handoff is required, thereby avoiding interruptions to the currently active conversation. The CT3's time slot runs at 32 Kbps and can be combined (8 slots) to achieve 256 Kbps for data applications.

2.6.2 DECT

Digital European Cordless Telephone (DECT) operates in the 1,880- to 1,900-MHz range (in Europe) and is the official third-generation European standard. It specifies 10 channels, each occupying a bandwidth of 1.728 MHz. Each of the channels is

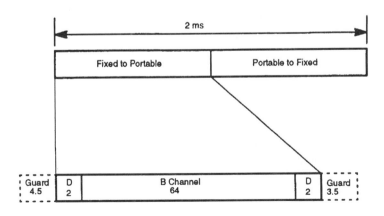

FIGURE 2.9 CT2 time slot and frame. (*Source:* [2].)

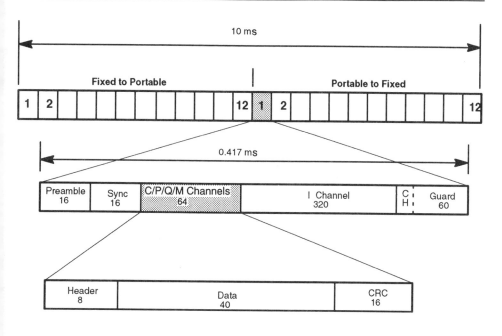

CRC: Cyclic Redundancy Test

CH: Parity Check Bits

Sync: Synchronization Sequence

FIGURE 2.10 DECT time slot and frame. (*Source:* [2].)

divided into 24 time slots, where two slots are allocated per voice conversation. The details of the frame format are shown in Figure 2.10.

The frame duration is 10 ms, and therefore each transmission burst lasts 10/24 = 0.417 ms. If we include the guard band of 64 bits, the 480 bits/slot translates to a bit rate of 1.152 Mbps. DECT employs GMSK as the modulation scheme with a wider bandwidth Gaussian filter (BT = 0.5) than that is used in GSM (BT = 0.3).

A summary of the standards for GSM, IS-54, CT2, and DECT is provided in Table 2.4.

TABLE 2.4
Summary of Standards

Parameter	GSM	IS-54	CT2
Forward band (MHz)	935–960	869–894	864–868
Reverse band (MHz)	890–915	824–849	864–868

TABLE 2.4 (continued)

Parameter	GSM	IS-54	CT2
Multiple access	TDMA	TDMA	FDMA
Duplex	FDD	FDD	TDD
Carrier spacing(kHz)	200	30	100
Channels/carrier	8	3	1
Bandwidth (kHz)	50	20	100
Channel rate (Kbps)	271	48.6	72
Modulation	GMSK	DQPSK	FSK
Mod efficiency	1.35	1.62	0.72
Voice rate (Kbps)	22.8	13	32
Control channel	SACCH	SACCH	D
Cntl chan. rate (bps)	967	600	200
Cntl chan. msg (bits)	184	65	64
Cntl chan. delay (ms)	480	240	32

Source: [2].

2.6.3 DCS-1800

Recognition of the overcrowding occurring at the GSM's 900-MHz band gave rise to the digital cellular system at 1,800 MHz (DCS-1800) standards. DCS-1800 specifies the radio and network aspects for personal communications network (PCN) services. Shorter range RF propagation at 1,800 MHz combined with lower powers (1W and 250 mW handsets) expected for the handsets required small cell implementations. With appropriate frequency reuse, the DCS-1800 contained the characteristics to realize high-capacity networks. The changes to the GSM standard for the DCS-1800 was restricted to the modifications required in the RF area (from 900-MHz operation to the 1,800-MHz operation).

Figures 2.11(a,b) show the incorporation of the micro and macro cells in the DCS-1800 network.

A collection of macrocells will cover the background traffic and microcells will be activated during the traffic peaks. Up to ten 200m microcells can be deployed.

A modified standard based on DCS-1800 for PCS systems to operate in the 1,900-MHz frequency range is being developed for application to the U.S. markets.

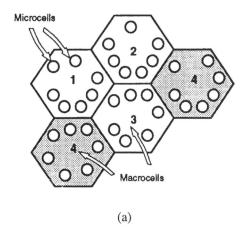

(a)

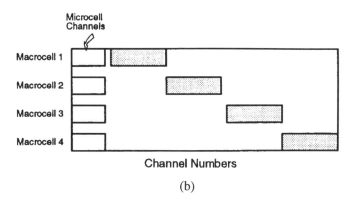

(b)

FIGURE 2.11 (a) Cell layout and (b) frequency plan for macro and micro cells for a DCS-1800 network. (*Source:* [6].)

2.6.4 N-CDMA and DS-SS

Qualcomm's core N-CDMA system uses a 14.4-Kbps air interface and 13-Kbps voice coders. The extended system proposed by Qualcomm uses a 76.8-Kbps air interface and works with higher quality codecs at 16 and 32 Kbps. Qualcomm is also planning to support these developments by building their own chip sets.

Omnipoint's DS-1900 system uses a combination of TDMA and direct sequence spread spectrum (DS-SS) to allow spectrum sharing and handset-initiated handoffs. The DS-1900 handsets will support data rates from 8 Kbps (voice) to 500 Kbps (data).

2.6.5 PCS Technology and Standards

Personal communication systems are expected to be provided by the cellular telephone carriers as they migrate to microcellular networks and also by ESMR providers such as Nextel as they expand their services beyond dispatching. Further, paging operators, as they expand from one-way to two-way paging, will become partners in the PCS framework, and so will be the licensees in the 1.8-GHz band as they develop different services on the new spectral framework.

The PCS band is a collection of smaller bands (most of them currently unused) and fall into two groups—the broadband and the narrowband. The associated frequency bands are shown in Tables 2.5 and 2.6.

TABLE 2.5
Narrowband PCS Channel Summary

Transmit/Receive Bandwidth	Number Available	License Area
50 kHz/50 kHz	5	Nationwide
50 kHz/12.5 kHz	3	Nationwide
50 kHz Unpaired	3	Nationwide
50 kHz/50 kHz	2	Regional
50 kHz/12.5 kHz	4	Regional
50 kHz/50 kHz	2	Major trading area
50 kHz/12.5 kHz	3	Major trading area
50 kHz Unpaired	2	Major trading area
50 kHz/12.5 kHz	2	Basic trading area

Source: [7].

TABLE 2.6
Broadband PCS Channel Summary

Frequency Block	Spectrum	Frequency Range	Geographic Scope
A	30 MHz	1850–1865/1930–1945 MHz	MTA
B	30 MHz	1870–1885/1950–1965 MHz	MTA
C	30 MHz	1895–1910/1975–1990 MHz	BTA
D	10 MHz	1865–1870/1945–1950 MHz	BTA

Frequency Block	Spectrum	Frequency Range	Geographic Scope
E	10 MHz	1885–1890/1965–1970 MHz	BTA
F	10 MHz	1890–1895/1970–1975 MHz	BTA
Unlicensed	20 MHz	1910–1930 MHz	Nationwide

Source: [7].

The narrowband spectrum will be primarily used for data services such as paging, messaging, and low-rate data transmission. Six licensed full-duplex frequency blocks that are part of the broadband group will be reserved for applications that require high bandwidths. The 20-MHz spectrum allocated for unlicensed PCS services will support low-power limited-range applications where the equipment will be owned and operated by the end users in their own premises. Wireless LANs and wireless PBXs will likely to be important applications for the unlicensed spectrum.

The Joint Technical Committee (JTC) formed by the members of the TIA TR46 group and the T1P1 group have come up with seven different air-interface standards for PCS, summarized in Figure 2.12.

The PCS service provider has the burden of choosing one or more of the above technologies as the infrastructure framework.

2.6.6 Microwave Relocation for PCS Deployment

PCS systems in the 1,850- to 1,990-MHz band will have to coexist with the existing microwave systems until they are upgraded or relocated. Fortunately, since the microwave links are stationary, the impact of these systems on others can be quantified if accurate configuration data is available. The estimate is that close to 90% of these offending microwave links are analog, many having been in operation for more than 30 years. Older technology with 3-dB IF filters several megahertz beyond channel edge are susceptible to significant adjacent channel interference. Interference mitigation can be achieved by using a mixture of techniques including base station siting, frequency engineering, sectoring cells, downtilting antennas, and power control (reduction).

2.7 MOBILE SATELLITE SERVICES

Mobile satellite services (MSS) are suited for providing services to large population of users located in a wide geographic area. MSS services are becoming increasingly popular because of the following reasons:

Parameter	Tag-1	Tag-2	Tag-3	Tag-4	Tag-5	Tag-6	Tag-7
Heritage	New	IS-95-based CDMA	PACS	IS-136-based TDMA	DCS-based TDMA	DCT-based TDMA	New
Access method	CDMA/TDMA/FDMA	DS-CDMA	TDM/TDMA	TDM/TDAM	TDMA	TDMA	DS-CDMA
Duplex method	TDD	FDD	FDD	FDD	FDD	TDD	FDD
Bandwidth	5 MHz	1.25 MHz	300 kHz	30 kHz	200 kHz	1.25 MHz	5 MHz
Bit rate (no overhead)	8 Kbps	8 and 13.3 Kbps	32 Kbps	7.95 Kbps	13 Kbps	32 Kbps	32 Kbps
Voice channels per carrier	32 (8 Kbps CELP)	20 (effective) + soft handoff	8	3	8	12	64
Modulation	QCPM	OQPSK/QPSK	$\pi/4$ D-QPSK	$\pi/4$ D-QPSK	GMSK	$\pi/4$ D-QPSK	QPSK
Error control (voice)	None	FEC	None	FEC	FEC	None	FEC
Frequency reuse (N)	3	1	16 × 1	7 × 3	7 × 1 and 3 × 3	9	1
Maximum average subscriber power	10 mW	200 mW	25 mW	200 mW	125 mW	20.8 mW	200 mW
Subscriber-unit power in time slot	1 W	–	200 mW	600 mW	1 W	250 mW	–
Time frame length	20 ms	20 ms	2.5 ms	40 ms	4.615 ms	10 ms	–
Time slot length	625 μs	–	312.5 μs	6.7 μs	577 μs	417 μs	–
End-to-end speech delay	80 ms	50 ms	9 ms	110 ms	90 ms	10 ms	9 ms
Equalizer	No	No	No	Yes	Yes	No	No
Vocoder	CELP (8 Kbps) used, ADPCM (16, 24, 32, 40 Kbps) possible	Variable-rate QCLP (8, 4, 2, and 1 Kbps) + 13.3 Kbps variable	ADPCM (32 Kbps)	VSELP (8 Kbps)	RPE-LTP (13 Kbps)	ADPCM (32 Kbps)	ADPCM (32 Kbps)

ADPCM = adaptive-differential pulse-code modulation; CDMA = code-division multiple access; CELP = code-excited linear predictive; DS-CDMA = direct sequence CDMA; DCS = digital cellular system (GSM-based); DCT = digital cordless telephone; D-QPSK = digital quadrature phase-shift keying; FDD = frequency division duplexing; FDMA = frequency-diviation multiple access; FEC = forward error correction; GMSK = Gaussian minimum shift keying; NA = not available; OQPSK = offset quadrature phase-shift keying; PACS = personal-access communication services; QCLP = quadrature codebook linear predictive; QCPM = quadrature continuous phase modulation; QPSK = quadrature phase-shift keying; RPE-LTP = regular pulse-excitation long-term predictor; TDD = time division duplexing; TDMA = time-division multiple access; VSELP = vector sum-excited linear predictive

FIGURE 2.12 Summary of air-interface standards for PCS. (*Source:* [7].)

- Advanced satellite antennas can project focused cellular patterns onto the Earth's surface, enabling frequency reuse.
- Satellites in low orbits allow the use of low-power handheld terminals.
- Digital multiplexing technologies allow hundreds of users to simultaneously access a single satellite transponder.

Further, the satellite communications together with the emerging personal communication technologies are creating synergistic services that offer the promise of instant worldwide communications.

Three different orbits can be used to deploy satellites, as shown in Figure 2.13.

- *Equatorial orbits.* Satellites in equatorial elliptical orbits circle the equator. In geosynchronous equatorial orbits where satellites are located 22,300 miles above the Earth, satellites and the Earth rotate at the same angular speed.

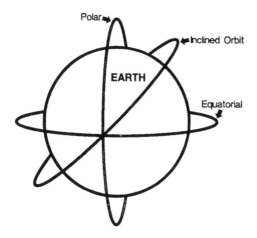

FIGURE 2.13 Satellite orbits.

Consequently, the satellite appears to be stationary with respect to the transmitters and receivers on Earth and essentially functions as a broadcast repeater.

- *Polar orbits.* These orbits revolve around the poles and are nonsynchronous.
- *Inclined orbits.* Elliptical inclined orbits require extensive tracking and are mainly used in military applications.

Access methods include FDMA, TDMA, and CDMA. Derivations from these generic access methods that are popular for satellite communications are single channel per carrier (SCPC) FDMA, which uses preassigned access, and demand assigned multiple access (DAMA). Satellite communications bands are listed in Table 2.7.

TABLE 2.7
Satellite Communication Bands

Frequency	Band	Uplink Range (GHz)	Downlink Range (GHz)	Use
6/4	C	5.925–6.425	3.7–4.2	Commercial
8/7	X	7.9–8.4	7.9–8.4	Military
14/11	Ku	14.0–14.5	11.7–12.2	Commercial
17/12		17.3–17.8	12.2–12.7	Direct broadcast
30/20	Ka	27.5–30.5	17.7–21.2	Commercial
30/20	Ka	30.0–31.0	20.2–21.2	Military

TABLE 2.7 (continued)

Frequency	Band	Uplink Range (GHz)	Downlink Range (GHz)	Use
44/20	Q	43.5–45.5	20.2–21.2	Military

Source: [8].

2.7.1 Geosynchronous Systems

Geosynchronous satellites (GEOS) require a costly delivery system ($50 to $100 million) to launch them into orbit and therefore are designed to have long life spans. The GEOS were originally placed 4 degrees apart in the geosynchronous orbit to prevent interference. The spacing of the satellites was later reduced to 2 degrees to relieve congestion in the space segment. Satellite supporting high-powered direct broadcast applications required 9 degrees of separation.

American Mobile Satellite Corporation (AMSC) has two satellites that provide circuit-switched voice, circuit-switched data, and packet-switched data services. Each satellite supports 4,200 channels of 5-kHz width. The mobile downlink will operate from 1,530 to 1,559 MHz and the uplink will from 1,631.5 to 1,660.5 MHz. The Earth station feeders will operate in the 10-GHz (downlink) and 13-GHz (uplink) bands.

Other GEOS services providers include Celsat, which plans to combine GEOS with terrestrial microcellular telephone network to provide nationwide coverage. The satellite creates more than 100 spot beams (supercells) and uses CDMA technology to support data transmission speeds of up to 144 Kbps. Video, fax, low-speed voice, and data services are expected to be provided.

The main disadvantage of the GEOS is that the propagation delay is in the order of quarter second, uplink and a downlink.

2.7.2 Low-Earth-Orbit Systems

The low-Earth-orbit systems (LEOS) are currently considered the most suitable approach for providing global communications as an important component of the PCS revolution. Several LEOS schemes are being investigated and funded, the most popular being Motorola's Iridium system. Since the satellites do not remain in fixed positions relative to Earth, a fleet of satellites with carefully defined orbital paths need to be deployed to provide continuous coverage to defined geographical areas on Earth. Low altitude enables the use of handsets/mobile terminals with low power, launching of satellites is less costly, and the propagation delay is significantly less (as low as 1/50th) than that with geosynchronous satellites.

The Iridium system uses 66 satellites with 11 satellites in each of 6 polar orbits. These satellites will be spaced 32.7 degrees apart and will move at approximately 16,700 miles per hour from north to south and 900 miles per hour westward at 423 nautical miles above the Earth. Each satellite projects 37 spot beams using frequencies in the L-band range. The center spot beam will be surrounded by three rings of beams of 6, 12, and 18 in number, each spot beam covering an area approximately 370 nautical miles in diameter. The combined beams cover an area of approximately 2,200 nautical miles in diameter. A satellite is visible to a stationary user for approximately 8–10 minutes. Unlike terrestrial cellular systems, the motion of the cells rather than the mobile users will be the predominant cause for handoffs.

An overview of an Iridium network is shown in Figure 2.14. In addition to the space segment, the network includes a gateway segment that comprises Earth stations that provide connectivity to the terrestrial public-switched telephone network (PSTN). Around 20 such gateways are planned for Iridium. When a satellite cannot access a gateway directly, it can pass the traffic via its side antenna to adjacent satellites via a trunk channel until the traffic can be downlinked to a gateway.

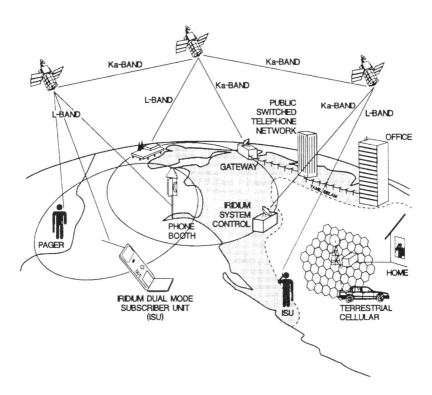

FIGURE 2.14 Iridium network overview.

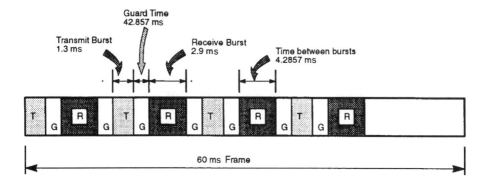

FIGURE 2.15 Iridium TDMA frame format. (*Source:* [8].)

As shown in Figure 2.14, the gateway feeder links (18.8–20.2 GHz range for downlink and 27.5–30.0 GHz range for uplink) and the intersatellite communication links (22.55–23.55 GHz range) will occur in the Ka band and will use bandwidths of up to 200 MHz.

Iridium uses a combined frequency and time division multiple access scheme for voice or data transfer. The TDMA frame is 60 ms long, and a 14-slot format allows each cell to be assigned two time slots. The frame format is shown in Figure 2.15.

The system uses differentially encoded, raised-cosine filtered QPSK as its digital modulation scheme.

2.8 WIRELESS LANs

Local area networking is another area that is being impacted by the wireless technology. The key element in wireless LANs is the elimination of hardwires, and other features such as mobility and handoff have not yet become critical to the LAN environment. The typical arrangement of the wireless LANs uses a point coordination node that provides coordination and control for all wireless stations within its coverage area, as shown in Figure 2.16.

While the IEEE 802.11 committee is working on finalizing the MAC layer and the supporting physical layer standards for wireless LANs, the physical layer implementations of wireless LAN products that have appeared in the market can be categorized as shown in Figure 2.17. Both infrared frequency (100 tetrahertz) and radiofrequency (RF) solutions are available. In the RF category, licensed microwave frequency range (18–23 GHz) or the unlicensed frequency range in the industrial, scientific, and medical (ISM) band (902- to 928-MHz, 2,400- to 2,478.5-MHz, and 5,725- to 5,850-MHz bands) are used.

For the ISM band, the maximum RF power is set to 1W by the FCC, although in practice it is limited to 100–200 mW. The range for these low-power RF system is

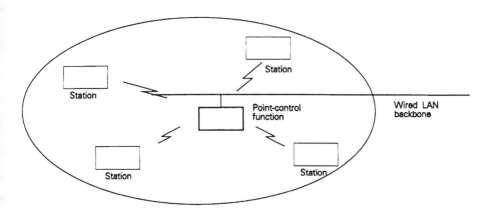

FIGURE 2.16 Wireless LAN configuration.

150 to 300 ft indoors and 800 ft outdoors. The ISM band LANs have to share the spectrum with others using the same band. The 900-MHz band can be used by cordless telephones, security systems, and baby monitors; PCS systems will be soon added to this list. The more popular 2,400-MHz band is also used by microwave ovens, diathermy machines, and licensed high-power users such as the police and railroad workers. Both direct sequence spread spectrum and frequency hopping spread spectrum technology have been used in the unlicensed band applications. AT&T's Wave-

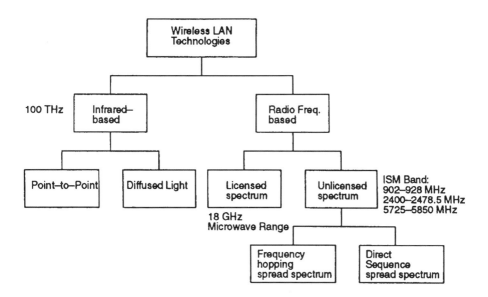

FIGURE 2.17 Wireless LAN technologies.

LAN uses direct sequence quadrature phase shift keying (QPSK) modulation and proprietary transmission protocols and can support speeds of up to 2 Mbps.

Infrared light systems are limited to 30–50 feet and the point-to-point service technique requires line of sight. The infrared cannot penetrate walls and floors. InfraLAN Technology's wireless LAN operates using this technique and can support a 4- to 16-Mbps token ring system or a 10-Mbps Ethernet LAN. Diffused infrared light technology (Photonics, Inc.) where diffused light provides a point-to-multipoint service is also available for wireless LANs.

Motorola's ALTAIR is a LAN offering based on the fixed-frequency licensed microwave technology. Operating at 18 GHz, the ALTAIR transports Ethernet frames (10 Mbps) using a CSMA/CA technology.

2.9 SUMMARY

The array of available wireless technologies and the multitude of ways products based on these technologies can be organized as competing and/or complementing infrastructures to support mobile computing systems is evident from the above overview. The wireless computing market and the technologies sustaining it are highly dynamic with a myriad of options and solutions, particularly so in the wireless data area.

CDPD promises to be an inexpensive technology, as the infrastructure cost is minimal and no CDPD-specific new RF spectrum is used. The throughput, unless the channels are heavily shared, is adequate for general use other than for real-time video or for running applications that involve high-resolution graphics. The throughput per channel (19.2 Kbps) is faster than the packet radio (2.4 to 14.4 Kbps), circuit-switched cellular (1.2 to 9.6 Kbps), or ESMR (4.8 Kbps), making CDPD an attractive choice for "fast" responses and data transfers.

Since the CDPD protocol is designed to mesh seamlessly with the existing IP-based networks, it can be used as transport for LAN-based wireless applications. Further, a wide spectrum of users use the Internet for commercial, personal, and also for general entertainment purposes. Since CDPD functions as an extended component of the extensive and heavily utilized Internet network, it is attractive to most users of the Internet.

Until the ubiquitous availability of data-capable technologies such as PCS become a reality, CDPD and other packet technologies will bridge the interim need, possibly interoperating among themselves to provide the users with wider coverage than any alternate single technology solution. It is also conceivable that if innovative adaptations are made to these existing technologies, they may be able to continue to hold the market share with the installed base of applications (if the numbers are large) for a longer period even after the arrival of new data technologies that support speeds demanded by bandwidth-hungry users.

References

[1]   Lee, W. C. Y., *Mobile Cellular Telecommunications Systems*, McGraw-Hill Book Company, 1989.

[2]   Goodman, D. J., "Trends in Cellular and Cordless Communications," *IEEE Communications*, Vol. 29, No. 6, June 1991, pp. 31–40.

[3]   Kay, S., "E-TDMA," *Cellular Business*, June 1992, pp. 60–65.

[4]   Brodsky, I., *Wireless: The Revolution in Personal Communications*, Norwood, MA: Artech House, 1995.

[5]   Mouly, M., and M. B. Pautet, "The GSM System for Mobile Communications," 1992.

[6]   Hadden, A,. "DCS-1800 - PCN In the Local Loop," *Telecommunications*, Nov. 1991.

[7]   Goldberg, L., "PCS: Technology with Fractured Standards," *Electronic Design*, Feb. 1995, pp. 65–78.

[8]   Bates, B., *Wireless Networked Communications*, McGraw Hill, 1994.

Bibliography

Cox, D., "Universal Digital Portable Radio Communications," *Proceedings of the IEEE*, Vol. 75, No. 4, April 1987, pp. 436–477.

Cox, D., "Wireless Personal Communications: What is it?" *IEEE Personal Communications Magazine*, Vol. 2, No. 2, April 1995.

Devoney, C., "The Unwired Nation," *Windows Sources*, April 1994, pp. 106–138.

Feher, K., *Wireless Digital Communications*, Englewood Cliffs, NJ: Prentice Hall, 1995.

Friedan, R., "Satellite-Based Personal Communication Services," *Telecommunications*, Dec. 1993 pp. 25–32.

Gibson, H. M., "Microwave Relocation and its Impact on PCS Deployment," *PCIA Journal*, May 1995.

Goldberg, L., "MAC Protocols: The Key to Robust Wireless Systems," *Electronic Design*, June 13, 1994, pp. 63–74.

Goldberg, L., "Satellite Communications: New Horizons," *Electronic Design*, Dec. 4, 1995, pp. 46–62.

Goodman, D. J., "Second Generation Wireless Information Networks," *IEEE Transactions on Vehicular Technology*, Vol. 40, No. 2 May 1991, pp. 366–374.

Gronert, E., and P. Heywood, "GSM: A Wireless Cure for Cross-Border Data Chaos," *Data Communications*, March 21, 1995.

Harte, L., *Dual Mode Cellular*, P.T. Steiner Publishing Co., 1992.

Kostedt, F., and J. C. Kemerling, "Practical GMSK Data Transmission," *Wireless Design and Development*, 1995, pp. 21–24.

Lee, W. C. Y., *Mobile Communications Design Fundamentals*, Wiley Interscience Publications, 1993.

Macario, R. C. V., *Cellular Radio: Principles and Design*, McGraw Hill Publications, 1993.

MacDonald, V. H., "The Cellular Concept," *The Bell System Technical Journal*, Jan. 1979, pp. 15–41.

Padgett, J. E., C. G. Gunther, and T. Hattori, "Overview of Wireless Personal Communications," *IEEE Communications Magazine*, Jan. 1995, pp. 28–41.

Rahnema, M., "Overview of the GSM System and Protocol Architecture," *IEEE Communications Magazine*, April 1993.

Seybold, A. M., *Using Wireless Communications in Business*, Van Nostrand Reinhold, 1994.

Viterbi, A. J., *CDMA: Principles of Spread Spectrum Communications*, Reading, MA: Addison Wesley Publishing, 1995.

CHAPTER 3
▼▼▼

ARCHITECTURE

The cellular digital packet data (CDPD) network is designed to operate as a peer multiprotocol, connectionless network that can operate as an extension of the existing data communications networks.

This chapter aims to provide an overview of the architecture of the CDPD network. This will be done using three different perspectives, as follows:

- *The service provider's view of the network.* This involves the various physical and logical network elements, and the interelement interfaces that go into providing the complete suite of network services.
- *The communications architecture.* This will describe the layers of the protocols at each of the significant network elements.
- *The subscriber's view of the network.*

3.1 SERVICE PROVIDER'S PERSPECTIVE OF THE CDPD NETWORK ARCHITECTURE

The CDPD network can be visualized as a collection of interconnected CDPD service provider networks. These CDPD service provider networks communicate among themselves and with other external networks over appropriate interfaces. Figure 3.1 shows a generic architecture of the CDPD network.

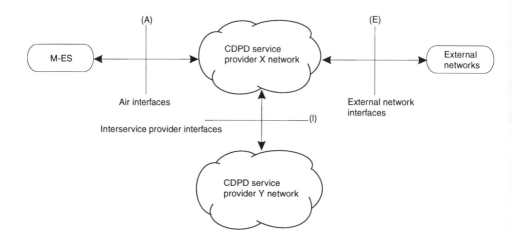

FIGURE 3.1 Interfaces between CDPD network components. (*Source:* [1].)

Figure 3.1 also shows the various interfaces (the A, E, and I interfaces) that will be discussed later in Section 3.8. This figure illustrates that a CDPD-compatible subscriber device (the M-ES) can interact with another external network (like the Internet) using one or more CDPD service provider networks.

The CDPD network is envisioned as an internetwork composed of multiple service provider domains. Each service provider domain can be constructed by interconnecting intermediate systems or routers. This view of the CDPD network is depicted in Figure 3.2.

Figure 3.3 shows the architecture within a single service provider's CDPD network. This figure illustrates the abstract functional network elements comprising this network. The network components are as follows:

- Intermediate system (IS) routers;
- Mobile data-intermediate system (MD-IS), which performs mobility management for the network;
- Mobile data base station (MDBS), which is the RF to landline relay;
- Mobile-end system (M-ES), which is the subscriber device;
- Fixed-end system (F-ES), which is comprised of the hosts attached to the landline network;
- Network servers, which are value-added network service application systems.

Figure 3.4 shows a reference model for the intercomponent interfaces in the form of the CDPD network reference model.

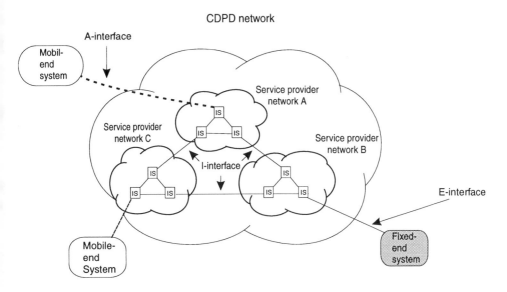

FIGURE 3.2 CDPD service provider domains and their interconnections. (*Source:* [1].)

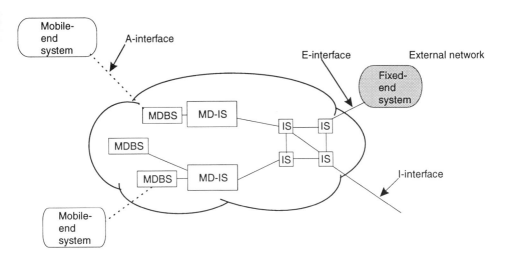

FIGURE 3.3 Single CDPD service provider's network. (*Source:* [1].)

The intercomponent interfaces are as follows:

- The airlink interface (the A interface);
- The external interface (the E interface);

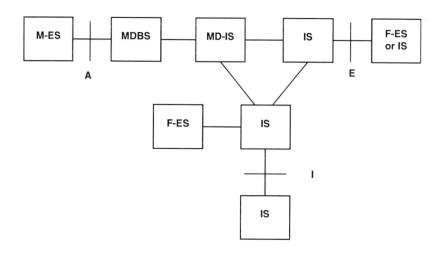

M-ES   - Mobile End System
MDBS - Mobile Data Base Station
MD-IS - Mobile Data Intermediate
           System
IS   - Intermediate System
F-ES   - Fixed End System

FIGURE 3.4   CDPD network reference model.

- The interservice provider interface (the I interface).

The network components and the intercomponent interfaces are described in the sections that follow.

3.2 INTERMEDIATE SYSTEMS

Intermediate systems allow for a network relay function that enables communication between any pair of end systems. The network relay function implies a system that receives datagrams from one correspondent network entity and forwards it to another correspondent network entity. Real physical equipment that can be deployed to serve this function uses standard commercial off-the-shelf routers that support Internet and OSI connectionless network service. Most real-world routers are unaware of mobility. ISs implement the network layer functionality (as also functions allocated to the layers below the network layer) as defined by the ISO reference model

[2,3]. The network layer must find a path through a series of connected ISs until the desired end system is reached. Routers along the path must forward network layer packets (datagrams) in the appropriate direction. They must deal with route calculation, fragmentation, reassembly, and congestion mitigation.

In the CDPD network architecture, ISs route connectionless network protocol (CLNP) [4] datagrams between MD-ISs and Internet protocol (IP) [5] datagrams between MD-ISs and F-ESs. ISs may also be used to route data between an MD-IS and its connected MDBSs.

The CDPD networks operate as a wireless extension of the existing connectionless networks. In the OSI network layer routing topology, a CDPD service provider network may be seen as a separate administrative routing domain.

3.3 MOBILE DATA INTERMEDIATE SYSTEM

The MD-IS performs routing functions based on the additional knowledge of the current location of the M-ES. In the CDPD network, the MD-ISs are the only network relay systems (the ISs or routers) that are aware of mobility. Location information is exchanged between MD-ISs using the mobile network location protocol (MNLP).

The MD-IS performs two distinct *mobility-aware* routing functions, which cooperate to provide location-independent network service. These functions are the mobile home function (MHF) and the mobile serving function (MSF).

The MHF relies on the concept that every M-ES is logically a member of a fixed home area. The home area provides the *anchor* or mobility-independent routing destination area for the ISs and ESs that are not mobile-aware. The MHF in the home area MD-IS operates a packet forwarding service (only for the forward direction) and maintains a database of the current serving area for each of its *homed* M-ESs. Packets received for any of its homed M-ESs are forwarded to the MSF of each serving area that the M-ES visits.

The MSF of an MD-IS operates a routing service for all visiting M-ESs in its serving area. The MSF relies on the mobile network registration protocol (MNRP) that is used by an M-ES to request access to the network. When the M-ES sends a registration request for a network entity identifier (NEI)—explained in Section 3.5—the MSF notifies the home MD-IS (the MHF functional module) of the M-ESs current location. The MSF also performs the function of routing data packets for M-ESs within its area toward the current subnetwork point of attachment based on local knowledge of the subscriber's current cell. The mobile data link between each M-ES and the serving MD-IS is controlled by the MSF, and this function performs the mobility management for M-ESs within the coverage area of the serving MD-IS.

3.4 MOBILE DATA BASE STATION

Each cell needs a logical MDBS function. A cell may operate multiple CDPD channel streams, each requiring a logical MDBS relay function. The MDBS performs the layer 2 data link or media access relay functions for a set of physical radio channels serving a cell. In the reverse direction (from the M-ES to the network), the M-ESs communicate over the radio interface to the MDBS, which then relays the information over a landline interface to the connected MD-IS. In the forward direction, the MDBS receives layer 2 datalink information from the MD-IS over a landline data network and then relays it over the airlink over physical RF channels.

The MDBS is responsible for control of the radio interface of all its connected channel streams. In a typical CDPD service provider network, an MDBS may be configured to control just one sector of a cell with multiple channel streams, or it may be set up to control all the three sectors of a trisectored cell site.

3.4.1 Channel Stream

A channel stream is a bidirectional communications path between an MDBS and a group of M-ESs, using a single RF channel pair at a time within a single cell. The channel stream can be thought of as a pair of point-to-multipoint channels. In the forward direction, the MDBS transmissions are received by all M-ESs listening to that channel stream. In the reverse direction, the MDBS receives the transmissions of the M-ES that is transmitting at that time. The reverse channel operates as a time-domain multiplexed channel. An M-ES is not expected to receive the transmissions of another M-ES directly. The RF channel pair used by the channel stream changes over time as a result of channel hopping to avoid interference or collision with cellular voice users.

3.5 MOBILE-END SYSTEM

The M-ES is the network component by which CDPD network subscribers gain access to the CDPD network. The mobile-end system uses the wireless interface to communicate with another end system (the other end system may be mobile or fixed.) As the M-ESs are potentially mobile and their physical location may change with time, continuous network access has to be maintained. This communication between the M-ESs and the CDPD network is done using the A interface. The CDPD network ensures that datagrams addressed to an M-ES will continue to reach the M-ES even as its physical location changes. The network tracks the location of the M-ESs and routes network layer datagrams to them and from them accordingly.

The M-ES provides the user of the CDPD network with application services, support services, and management services.

The application functions resident on the M-ES may be endpoint functions, such as providing wireless access for a virtual terminal application, or it may perform intermediate relay functions, such as providing network connectivity to allow remote end systems to access CDPD via a private LAN. The M-ES applications make use of standard protocols where possible.

Support services provided by the M-ES include transmission and reception of data across the airlink on either a full-duplex or half-duplex basis, medium access control to the airlink, mobile data link procedures, subnetwork convergence features, and end-to-end delivery of network level packets that are either based on IP or CLNP.

Management support is provided by the M-ES for control and configuration of the M-ES support services. These include radio resource management, security services, and registration services.

3.5.1 Network Entity Identifier

The M-ES is addressed by its network entity identifier (NEI). This identifier depends on the particular network layer protocol entity used by the M-ES to exchange network layer protocol data units with the MD-IS. The CDPD specification allows for the use of OSI CLNS and IP as the network layer protocols. The NEI typically consists of a network id followed by a host id.

In the CDPD networks deployed as of the writing of this book, the NEIs used by the service providers are IP addresses assigned by inter-NIC.

3.5.2 M-ES Internal Structure

The M-ES may be functionally decomposed into three subsystems:

- *The mobile application subsystem (MAS).* This contains those modules that are independent of the CDPD environment, such as the application software and the transport level software. The entities contained in the MAS are the subscriber applications (SAs).
- *The subscriber unit (SU).* This is the subsystem that supports the CDPD interfaces and functionality. It may be an integral part of the M-ES or it may be a separable physical component.
- *The subscriber identity module (SIM).* The SIM subsystem defines the identities and access rights of the M-ES's user(s). The SIM may be a subsystem of the SU or a subsystem of the M-ES, or it may be an integral component of one or the other subsystem.

The functional decomposition is illustrated in Figure 3.5.

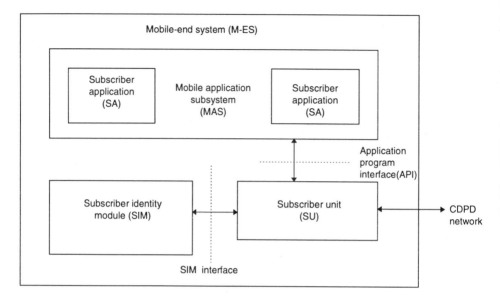

FIGURE 3.5 M-ES functional architecture.

The interface between the SU and the MAS is called the application program interface (API). The interface between the SIM and the SU is called the SIM interface (SIMI).

The particular application's environment and the design constraints of a particular manufacturer of the M-ES device will decide the API and the SIMI. Separate protocols and protocol profiles may be required to support communication between the MAS and the SU and between the SU and the SIM.

3.6 FIXED-END SYSTEM

Fixed-end system (F-ES) is a generic term applied to a non-mobile-end system in the OSI terminology. This is typically a host computing platform connected over a land-line data network to the MD-IS. Applications within the M-ES communicate with applications in F-ESs, utilizing the CDPD network as a communication backbone.

F-ESs may be external or internal, based on whether they provide external application services or application services internal to the CDPD network.

An external F-ES is owned, operated, administered, and maintained outside the direct control of CDPD network service provider. An example of an external F-ES is any host platform running a corporate database server or an e-mail server in an office LAN environment. A virtual terminal gateway and directory services are other examples of application services running on external F-ESs.

The term *internal F-ES* is sometimes applied to value-added network service application systems like, for example, a network management system (NMS). These systems are administered and maintained by the CDPD network service provider. Not all of these end systems are visible or directly addressable by CDPD network subscribers. Some other examples of internal F-ESs would be the location server, the domain name server, the accounting server, the message transfer server, and the authentication server on the CDPD network.

3.7 NETWORK SERVERS

Figure 3.6 shows the major components of the network servers. The network servers operate the following network services using internal F-ESs:

- Authentication;
- Accounting;
- Network management;
- Message transfer service;
- Location service.

3.7.1 Authentication Server

Because the CDPD network is a public wireless data communications service, it is vulnerable to fraudulent usage. To minimize the possibility of such unauthorized us-

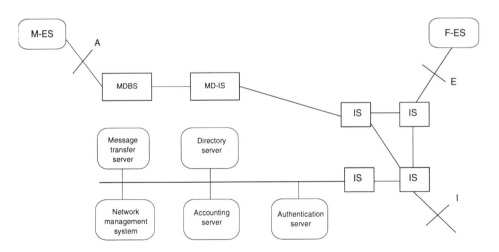

FIGURE 3.6 Network servers.

age, an M-ES authentication mechanism is defined for the CDPD network. The authentication server in conjunction with the home MD-IS provides a mechanism that allows an authorized M-ES to gain access into the CDPD service provider network.

3.7.2 Accounting Server

The accounting server network service application provides a secure, reliable repository for information collected from the accounting meters resident at the MD-IS. It utilizes a CCITT X.400 message transfer system for delivery, security, and temporary storage for later delivery.

The accounting meters in the connected MD-ISs deliver traffic matrix segments to serving accounting distributors (SAD). The SADs sort information obtained from the accounting meters and send this out as home Accounting segments to multiple home accounting collectors (HAC) and serving accounting collectors (SAC). Consolidation accounting collectors consolidate the accounting information collected for several subscribers that may all belong to the same accounting group.

The model of operation for the accounting service model is illustrated in Figure 3.7.

3.7.3 Network Management Server

This manages the various physical and logical elements making up a CDPD service provider network. The common management information protocol (CMIP) [6] and common management information services (CMIS) are used to manage the network. The simple network management protocol (SNMP) [7–9] is used to manage M-ESs.

The X.700 model of management is based on the idea of making visible across the management interface an abstraction of the resource being managed. An agent role is defined as that of the managed object to be operated on by the network management system. The manager role is defined as that of the system that will operate on the managed object.

The network management system interfaces, including the agent and manager roles for the various network elements, are shown in Figure 3.8.

3.7.4 Message Transfer Server

This provides a generic message store and forward service that is utilized by other CDPD network service application services like the accounting service and the location service. The message transfer service is based on the 1988 CCITT X.400 series of recommendations.

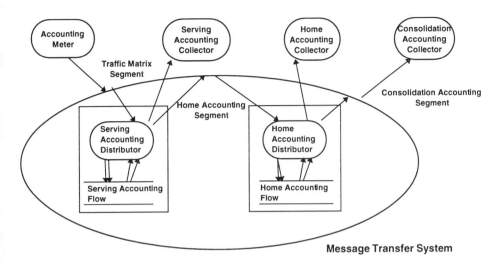

FIGURE 3.7 Accounting service model of operation.

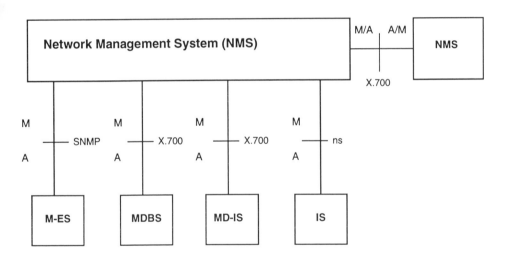

M - Manager Role
A - Agent Role
M/A - Both Manager and Agent Role
A/M - Both Agent and Manager Role
ns   - Not Specified

FIGURE 3.8 Network management interfaces.

3.7.5 Location Server

The location server runs the subscriber area location service that provides a report on the geographic location of subscribed M-ESs. Location reporting may be used by delivery services and fleet vehicle owners to track the location of their vehicles. This service is built upon a secure delivery mechanism to ensure that only authorized recipients are able to receive location information.

3.8 NETWORK INTERFACES

A CDPD service provider network must accommodate the interfaces to the subscribers (the A interface), to other service provider networks (the I interface), and to external networks and external F-ESs (the E interface).

Figure 3.9 illustrates the network interfaces from a service provider's perspective.

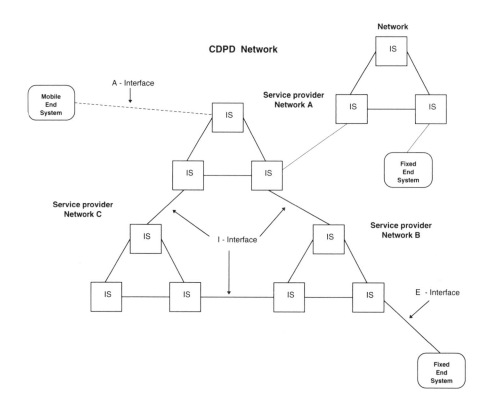

FIGURE 3.9 Network interfaces.

3.8.1 The A Interface

This describes the CDPD service provider's airlink interface for providing services over the RF airlink to subscribers. The CDPD subscribers use M-ESs to access network services through this airlink interface.

3.8.2 The E Interface

The external interface is the CDPD service provider's interface to external networks or external F-ESs. The external network may be a public data network like the Internet. The external application service providers communicate with CDPD subscribers through this interface.

3.8.3 The I Interface

This provides an interface between service providers to exchange network layer datagrams, mobility information, accounting data, subscriber profiles, and administrative information. This interservice provider interface allows support of CDPD network services across all areas served by CDPD service provider domains. This interface is not visible outside the CDPD network.

3.9 COMMUNICATIONS ARCHITECTURE

3.9.1 Reference Architecture

The CDPD Communications Reference Model uses the layering concept as defined in [10]. Since the layers are basically derived from the ISO OSI model, we do not intend to describe each of the layers and the services provided by each layer in this section. Figure 3.10 shows the CDPD Communications Reference Model with a brief description of each of the layers.

Each of the layers in the CDPD network communications architecture may be further partitioned into similar sequences of sublayers. Each layer or sublayer is defined with layer service access points, layer protocol, layer service primitives, and a layer management entity.

A layer service access point is the point at which layer N provides services to layer $N + 1$. As an example, a network layer service access point (NSAP) is the point at which the network layer provides services to the transport layer. A layer service access point is also the point at which layer N accesses services provided by layer $N - 1$. Taking the above example, the network layer communicates with the data link layer through the data link layer service access point (LSAP). Associated with each LSAP is one or more data link connection endpoints.

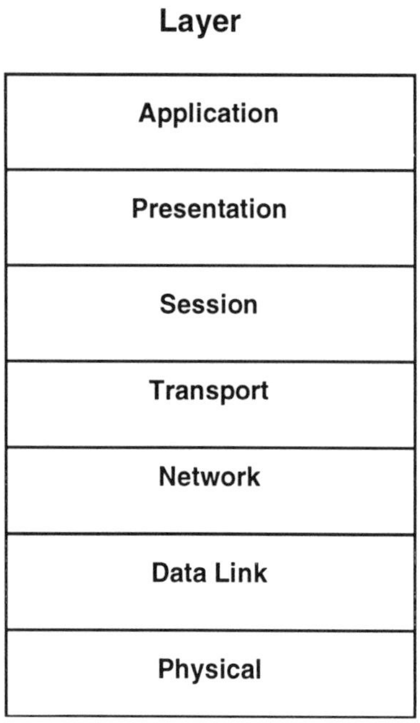

Layer

Application
Presentation
Session
Transport
Network
Data Link
Physical

FIGURE 3.10 CDPD Communications Reference Model.

Layer entities exist in each layer. Peer entities are entities in the same layer but in different systems. The layer protocol is a peer-to-peer protocol that provides the mechanism for data to be exchanged between peer layer entities.

Layer entities request services from a lower layer using service primitives. The primitives are an abstraction for the exchange of information and control between the layer entity and adjacent layer entities. However, implementation of individual layers in a communication subsystem may not necessarily use the same primitives.

Figure 3.11 shows the layered architecture concept that is used between a pair of communicating end systems. This figure illustrates the concept of peer-to-peer exchange of information.

3.9.2 CDPD Network Subprofiles

A particular CDPD network element or network service may be implemented by conforming to building blocks of protocol subprofiles that are defined by the CDPD specification [1]. These subprofiles are specific multilayer protocol definitions. One major objective of the CDPD network of achieving interoperability across all CDPD service providers is served by using a consistent subprofile to perform the functions of end-to-end reliable data transfer.

The CDPD specifications define three major classes of subprofiles:

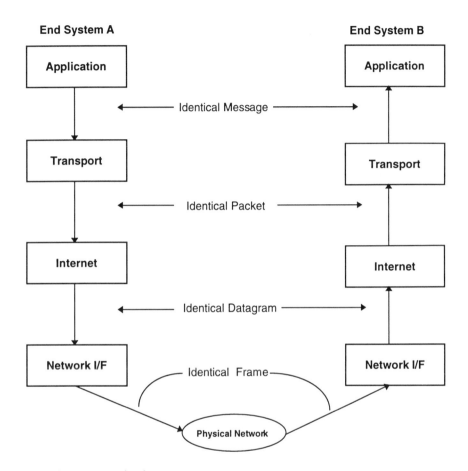

End System A

End System B

FIGURE 3.11 Layered architecture.

1. *Application subprofiles.* These are for the support of CDPD service provider application services like messaging, virtual terminal, directory services, and network management. Each application may require a different set of services from the application, presentation and session layers. The application subprofiles specify the layer 5, 6, and 7 requirements for each application service.

2. *Lower layer subprofiles.* Interoperable data transfer requires support of lower layer services provided by transport protocol class 4 (TP4) and the connectionless mode network protocol (CLNP). The use of TP4 and CLNP provides a common basis for reliable end-to-end communication across all types of subnetworking technologies. Appropriate subprofiles are defined for end systems, MD-IS, and intermediate systems.

3. *Subnetwork subprofiles.* Different technologies for local and wide area networking exist. Selection among these is typically by criteria of usage, performance, and cost. The variety of standard subnetwork technologies include X.25 WAN, frame relay, and point-to-point links.

This partitioning of the CDPD network services and protocols reflects the independence of user requirements for upper layer application services, lower layer communications services, and use of specific real and commercially available subnetwork technologies.

Options exist within a subprofile that relate to the support of optional functionality, the role in which a system performs. These options parametrize a given subprofile into CDPD-specific network element definitions that are supported by a carrier or service provider.

Airlink Protocol Profile

The airlink provides the wireless radiofrequency data communications between the M-ES and the MDBS and MD-IS. Using the functions provided by the CDPD airlink, the mobile serving function at the MD-IS communicates with each M-ES through the MDBS. The MDBS acts as a data link relay system.

The airlink interface provides access to the CDPD network using either the Internet protocol [5] or the ISO connectionless network protocol [4].

The CDPD airlink protocol profile is a collection of subprofiles that go into each of the network elements, namely, the M-ES, MDBS, and MD-IS. Figure 3.12 shows the airlink protocol profile.

The figure demonstrates that network layer protocol data units (NPDUs or packets) are transmitted across a mobile data link connection between the M-ES and the MD-IS using the mobile data link protocol(MDLP). The link layer PDUs (called MDLP frames) are sent out in two steps, namely exchanges between the MD-IS and MDBS and exchanges between the M-ES and MDBS. The MD-IS to MDBS communications are dictated by the particular subnetwork protocol in use. Communication between the M-ES and MDBS make use of the MAC and physical protocols that are discussed in subsequent chapters of this book.

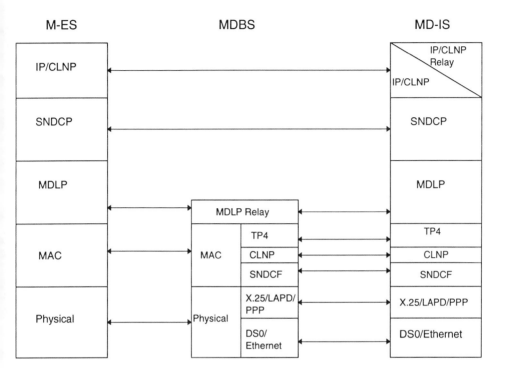

FIGURE 3.12 Airlink protocol profile.

References

[1] CDPD Specification, Version 1.1, CDPD Forum.

[2] ISO-7498: Information Processing Systems—Open Systems Interconnection: Basic Reference Model.

[3] Tannenbaum, A. S., *Computer Networks*, Englewood Cliffs, NJ: Prentice Hall, 1981.

[4] ISO 8473: Information Processing Systems—Data Communications Protocol for Providing the Connectionless Mode Network Service and Provision of Underlying Service.

[5] RFC 791: Postel, J. B., Internet Protocol, September 1981.

[6] CCITT-X.711: Information Technology—Open Systems Interconnection—Common Management Information Protocol Specification, Part 1.

[7] RFC 1155: Rose, M. T., and K. McCloghrie, Structure and Identification of Management Information for TCP/IP-based Internets.

[8] RFC 1157: Case, J. D., M. Fedor, M. L. Schoffstall, and C. Davin, Simple Network Management Protocol (SNMP), May 1990.

[9] RFC 1213: McCloghrie, K., and M. T. Rose, Management Information Base for Network Management of TCP/IP-Based Internets: MIB-II.

[10] ISO 7498: Information Processing Systems—Communications Reference Model.

CHAPTER 4
▼▼▼

AIRLINK INTERFACE

4.1 INTRODUCTION

Within the context of the CDPD system architecture, the airlink protocol is primarily concerned with communication at the physical and medium access control (MAC) layers of the protocol stack between mobile-end systems (M-ESs) and the mobile data base station (MDBS). The complete airlink protocol additionally involves the mobile data link protocol (MDLP) and the subnetwork-dependent convergence protocol (SNDCP) layer, as shown in Figure 4.1. The information transmission over the air represents the slowest and most error-prone segment of a CDPD end-to-end "connection" between an M-ES and a fixed-end system. The implementation architecture and recommended procedures, therefore, have significant impact in the operation of a CDPD network.

This chapter concentrates mainly on the physical and the MAC layer portions of the airlink and subsequent chapters deal with MDLP and the SNDCP portions of the protocol.

4.2 OVERVIEW OF DATA FLOW

In the air interface protocol stack, application data ready for transmission undergoes several transformations before appearing as *bits* on the physical transmission channels. A clear understanding of this path and the associated data mapping is necessary

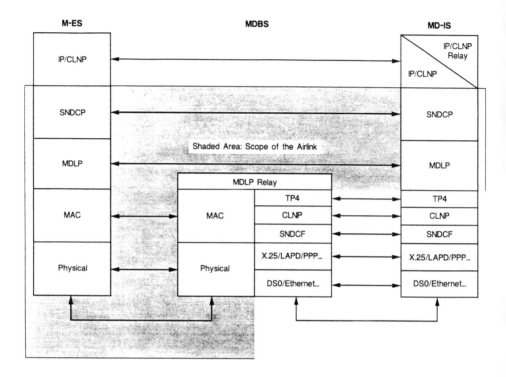

FIGURE 4.1 Airlink protocol stack. (*Source:* [1]. © CDPD Forum.)

to understand the functional details of the different layers and to form a coherent picture of the data transfer mechanisms in the CDPD system. Figure 4.2 lists the different processes involved. The data transformations at these different steps are summarized below.

- *Network layer (IP/CLNP).* The subnetwork-dependent convergence protocol (SNDCP) receives the network layer data packet.
- *SNDCP.* With SNDCP, optional header compression is implemented on the header portion of the IP or CLNP packet, and optional V.42 bis data compression is implemented on the data and header portion of the packet. The packet is segmented based on the maximum frame size handled by the link layer entity and an SNDCP header is added to each segment. Encryption is performed on the data portion (not the SNDCP header) and the segmented packet is forwarded to the MDLP entity.
- *MDLP.* The MDLP layer adds the MDLP header and forwards the frame to the MAC layer entity.
- *MAC.* With MAC, a sequence of frames is converted into a bitstream by inserting at least a single frame flag in between. The bit stream is blocked into con-

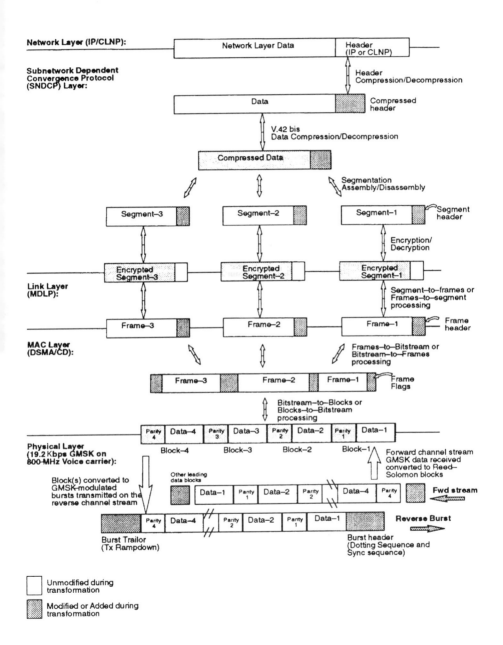

FIGURE 4.2 Data flow across the airlink stack.

secutive sets of bits and the RS encoding is performed. Information and Reed-Solomon (RS) code bits are exclusive-ORed with a pseudo-noise sequence (PN encoding).

- *Forward Channel.* The data and RS code bits (378 bits) are interleaved with an additional 42 bits containing forward synchronization word, decode status, and busy/idle flags to make up a 420-bit block.
- *Reverse Channel.* The data and RS code bits (378 bits) are interleaved with 7 continuity indicator bits to form 385-bit blocks. Multiple blocks (full-duplex M-ES) or a single block (half-duplex M-ES) is prefixed with a dotting sequence and a reverse synchronization word to form the reverse burst. Each bit is modulated as a symbol of a Gaussian minimum shift keying (GMSK) waveform.

These functions are performed in reverse when information bits are received from an RF channel. In the following sections, the functions performed by the physical and the MAC layers are described in detail.

4.3 MAIN FUNCTIONS OF THE LAYERS

The primary function of the physical layer is to transmit a sequence of bits as a modulated waveform and to receive modulated data, demodulate them, and generate a sequence of received bits. The 30-kHz bandwidth RF channels defined for the analog cellular system is used as the medium.

Major functions of the physical layer involve transmission and reception at the physical layer to air interface and the provision of interface primitives to control the operation of the physical layer at the physical/MAC layer interface. These can be summarized as follows:

- Tuning of transmit and receive frequencies, complying with the bandwidth requirements of the RF channels and the interface primitives providing the means to tuning the frequencies. The transmission at the M-ESs will be in bursts whereas the transmission at the MDBS will be a continuous bit stream.
- Provision of interface primitives, and the means to vary transmitted power levels and to measure the received power levels. Power conservation primitives also provide means to conserving battery power.
- Provision of interface primitives and means for, modulating and transmitting individual bits and receiving and demodulating individual bits.

The MAC layer protocol between the M-ES and the MDBS is intended to provide an efficient and error-free medium for data transfer. The MAC protocol defines the following:

- Mapping link layer frames passed from the link layer to Reed-Solomon blocks and constructing link-level frames from received RS blocks;
- Forward error-correction coding to correct or to detect uncorrectable blocks;
- Performing other formatting operations to generate blocks ready for transmission;
- Access signaling procedures that use slotted nonpersistent digital sense multiple access protocol with collision detection (DSMA/CD);
- Generating synchronization signaling to allow M-ES and MDBS to synchronize;
- Providing means for identifying the MDBS an M-ES is communicating with and also to detect co-channel interference.

Implementation considerations in both the MDBS and M-ES typically dictate the use of 16-bit/32-bit DSP(s) for MAC layer processing.

4.4 THE PHYSICAL LAYER FUNCTIONS

Tuning to a specific RF channel, controlling the output power, transferring and receiving data to/from the tuned RF-channel are the main functions performed by the physical layer.

4.4.1 RF Channel Management

Communication between the peer physical layer entities in an M-ES and MDBS takes place over a pair of RF channels provided for the operation of the analog mobile phone system (AMPS) and defined in [2]. There are 832 such RF channel pairs, defined as shown in Figure 4.3 and described below:

- A total of 50 MHz of spectrum is divided into two 25-MHz blocks separated by 45 MHz. The channel spacing is 30 kHz.
- Each operator within a geographical area is allocated half the available channels (416). The B frequencies are reserved for the wireline carrier and the A frequencies for the nonwireline operator.
- MDBS receive channel (reverse channel) at 825.030 MHz and the corresponding M-ES receive channel (forward channel) at 870.030 MHz (a 45 MHz separation) is defined as channel 1.

The physical layer will have the ability to tune to a specific pair of RF channels denoted by channel number for transmission and reception of bits between the M-ES and the MDBS.

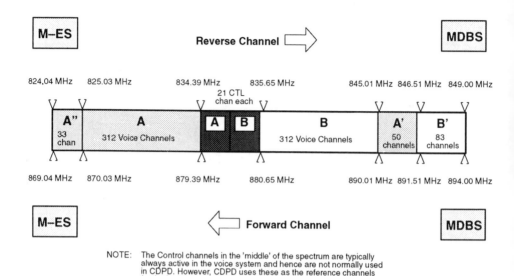

FIGURE 4.3 Spectrum allocation for AMPS/CDPD channels.

4.4.2 Data Transfer

Once a frequency is tuned, the physical layer must be capable of transmitting and receiving bits at the rate of 19.2 Kbps with an accuracy of 10 parts per million. The forward channel (the MDBS transmission) will be a continuous stream of data whereas the data transmission in the reverse channel will be in bursts by one or more M-ESs. In cases where there are no M-ESs using a channel, there may be no transmissions in the reverse channel.

The MDBS is designed for full-duplex operation, whereas two types of M-ESs, half duplex and full duplex, can be used with the CDPD network.

4.4.3 Transmit Power Requirements

The physical layer has the ability to set the power level to be used for transmission between an MDBS an M-ES. In addition, it has the ability to measure the signal level of received bits. Optionally, the physical layer in an M-ES may have the ability to suspend and resume monitoring of received channel as part of sleep mode management procedure, a mechanism designed to promote conservation of power in an M-ES.

The CDPD specifications define minimum performance for the M-ES and MDBS transmitters in terms of frequency stability, allowed phase noise, channel switching times, power stability, modulation requirements, and acceptable spurious emissions. In addition, for the M-ESs, four different device classes are defined in Table 4.1 with the following maximum nominal effective radiated powers (ERPs). The ERP is the combined effect of the mean output power level and the power gain or loss of the antenna system to be installed with the mobile-end system.

TABLE 4.1
Classes of M-ESs

Class	Maximum ERP	Range
I	6 dBW = 4.0W	–22 dBW to 6 dBW
II	2 dBW = 1.6W	–22 dBW to 2 dBW
III	–2 dBW = 0.6W	–22 dBW to –2 dBW
IV	–2 dBW = 0.6W	–34 dBW ±9 dB to –2 dBW

The MDBS transmit power, however, has no such limitations. The RF engineering requirements established for the carrier's network determine the transmit power at the MDBS. The site commissioning procedure will include the power output requirements. In typical cases, where the CDPD and AMPS cells sites are installed on a one-on-one basis, the CDPD power output is set to be the same as the AMPS transmit power.

4.4.4 Modulation/Demodulation

CDPD uses GMSK as the modulation scheme. This is a constant envelope modulation scheme and has efficient bandwidth characteristics. The parameters used for CDPD are:

- Modulation index $h = 0.5$, with a bit value 1 representing an instantaneous value of frequency greater than the carrier frequency. The modulation index is not to vary more than 1%.
- The premodulation Gaussian filter will have a bandwidth-data rate product of $BT = 0.5$, where T is the bit period and B is the filter bandwidth.

Technical details of the modulation scheme are further described in Section 4.5.

4.5 MODULATION TECHNIQUE

In communication systems employing radiofrequency channels, the two primary resources, transmitted power and channel bandwidth will be used as efficiently as possible. In power-limited channels, coding schemes will attempt to save power at the expense of bandwidth; and in bandwidth-limited channels, efficient modulation schemes will be used to obtain maximum data rates with acceptable bit error rates (BERs) within the allowed bandwidth.

4.5.1 Spectral Efficiency Considerations

CDPD airlink is essentially a bandwidth-limited channel and each direction of an RF channel has a carrier frequency in the 800–900 MHz range, each with a bandwidth of 30 kHz. They are also power limited to allow viable frequency reuse options. The M-ESs additionally have limited power transmission capabilities and therefore the reverse channels are power-limited. The MDBS transmissions further have to comply with the following restrictions on the emissions outside the 30-kHz bandwidth of a given channel:

- Emission in immediately adjacent channels centered ±30 kHz from the center frequency cannot exceed 26 dB below the mean output power level.
- Emission in the next adjacent channels centered ±60 kHz from the center frequency cannot exceed 45 dB below the mean output power level.
- Emission in any other channels centered ±90 kHz or more from the center frequency cannot exceed 60 dB below the mean output power or –13 dBm (50 microwatts), whichever is less.

The objectives of a spectrally efficient modulation schemes are twofold:

- Maximization of data rate per hertz of channel bandwidth, defined as bandwidth efficiency (bits/sec/Hz);
- Minimizing the signal power expended to achieve the above data rate at a given average bit error rate.

Further, when nonlinearity is present in channels, amplitude fluctuations in the signal generate extraneous sidebands that could introduce out-of-band interference. In CDPD, use of class-C power amplifiers to obtain maximum amplifier efficiency to conserve power introduces nonlinearity in the channels. Constant envelope modulation schemes do not generate such sidebands and therefore form a useful subset that can be used in nonlinear transmission channels.

Gaussian minimum shift keying, a constant envelope modulation scheme used in CDPD, satisfies the above requirements. This modulation scheme is also employed in other applications such as GSM.

4.5.2 Constant Envelope Modulation Schemes

Frequency shift keying (FSK) and phase shift keying (PSK) are two commonly known signaling techniques that possess constant envelope characteristics. If we consider binary communication at a signaling rate of $1/T$ pulses/sec, binary FSK refers to transmitting a carrier at one of two frequencies based on the value of the input pulse. In binary PSK, two opposite phases of the carrier are used, as shown in Figure 4.4. Bit error rate performance is a quantity that can be used to consider the effectiveness of different modulation schemes. It refers to the average number of errors in transmitting a bit stream through an ideal channel, a linear all-pass channel corrupted by additive white Gaussian noise (AWGN) with a power spectral density of N W/Hz. The plot of the ratio E/N, where E is the signal energy per information bit required to achieve different BERs can be used for comparisons.

Optimum receivers in such channels require matched filters and use recovered carrier phase reference. These are called coherent receivers and often provide the best BER values for antipodal (same shape but opposite polarity) signals. But noncoherent detection schemes often allow simple receiver implementations, but require noncoherently orthogonal signals. For example, for noncoherent demodulation of FSK signals, the two frequencies must be separated by a minimum of $1/T$, the signaling rate for noncoherent orthogonality. Even if coherent detection of FSK is used, it is still poorer by 3 dB in terms of E/N than PSK. But it can be shown that MSK, which is a coherently orthogonal FSK modulation scheme, requires only $1/(2T)$ Hz frequency separation and achieves performance equivalent to that of BPSK if a coherent receiver is used to base its decision after two bit periods. Further, the MSK can also be viewed as a particular variation of offset quadrature phase shift keying (OQPSK).

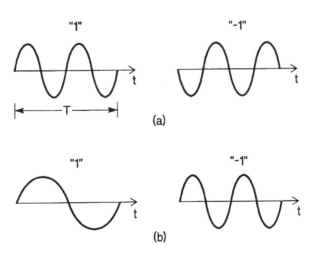

FIGURE 4.4 Basic (a) PSK and (b) FSK waveforms.

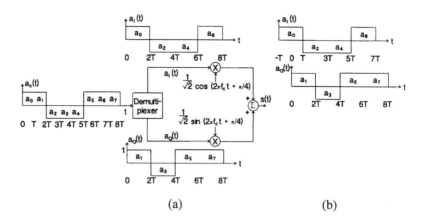

FIGURE 4.5 (a) QPSK modulator, and (b) data streams in OQPSK. (*Source:* [3]. © 1979 IEEE.)

4.5.3 QPSK and OQPSK

Bandwidth efficiency of binary PSK can be increased by two by using two carrier signals in quadrature, modulated with two bit streams derived from the input bit stream, as shown in Figure 4.5. Denoting $a_I(t)$ and $a_Q(t)$ as two bit streams consisting of even and odd bits from the input stream, the resulting modulated QPSK waveform is shown in Figure 4.6. The odd and even bit streams transmitted at the rate of $1/(2T)$ baud are synchronously aligned in QPSK such that their transitions coincide. Offset

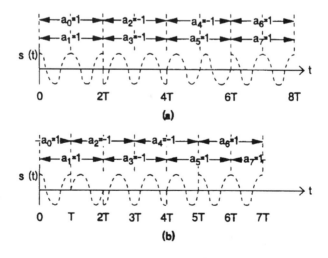

FIGURE 4.6 (a) QPSK waveforms, and (b) OQPSK waveforms. (*Source:* [3]. © 1979 IEEE.)

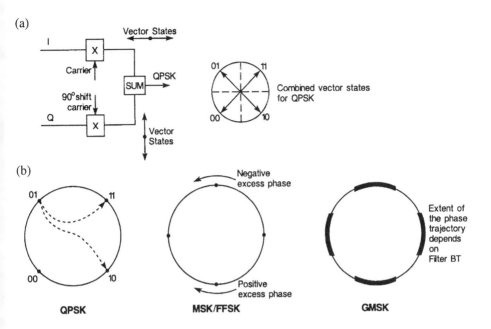

FIGURE 4.7 Vector state diagrams.

QPSK (OQPSK) modulation is obtained by having an offset in the relative alignments of the two data streams by an amount equal to T.

The power spectral density of QPSK and the OQPSK remains the same, having the shape $(\sin 2\pi fT/(2\pi fT))^2$ associated with the rectangular pulse used for signaling. In QPSK, the carrier phase can change only once every $2T$. Change can be ±90 if only one component changes and a phase shift of 180 if both components change. In OQPSK, due to the shift, only one component can switch at a time. Therefore, phase changes are limited to 0, +90, or −90 every T seconds.

A vector state diagram can be used to illustrate where the phase of the modulated signal is at the center of each symbol period. For QPSK, if there is no coherence between the data rate and the carrier frequency, the waveform may be constrained to move across the phase circle or rapidly round the circle as shown in Figure 4.7(a).

4.5.4 Minimum Shift Keying

Minimum shift keying can be thought of as a special case of OQPSK with sinusoidal pulse weighting. If sinusoidal pulses are employed instead of rectangular shapes, the modified signal can be represented as

$$s(t) = a_I(t) \cos(\pi t/2T)\cos 2\pi f_c t + a_Q(t)\sin(\pi t/2T)\sin 2\pi f_c t$$

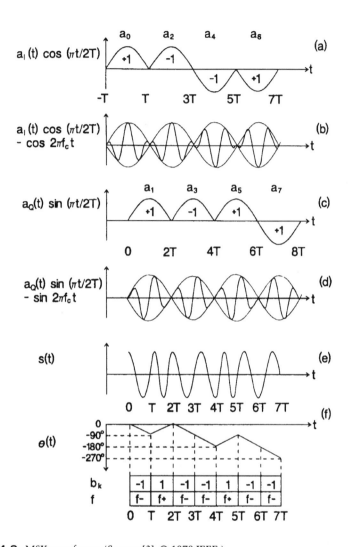

FIGURE 4.8 MSK waveforms. (*Source:* [3]. © 1979 IEEE.)

Figure 4.8 shows the various components of the MSK signal defined by the above equation. The following properties of MSK can be deduced from these points:

- MSK has constant envelope;
- Phase continuity at the bit transition points;
- The resulting signal is an FSK signal with signaling frequencies of f+ = fc + 0.25(1/T) and $f-$ = f_c − 0.25(1/T). That is, the frequency deviation is half the bit rate.

In MSK, the constant envelope with the phase continuity at bit transitions results in the carrier moving from one phase state to the next around the phase circle, as shown in Figure 4.7(b).

4.5.5 Gaussian Minimum Shift Keying

MSK still suffers from poor adjacent channel sidebands arising from its FSK-type spectrum, as shown in Figure 4.9. Considerations to limit the effect of the rapid change in frequency that occurs in the MSK scheme gave rise to the GMSK.

The output power spectrum of MSK can be easily manipulated by using a premodulation lowpass filter keeping the constant envelope property. To make the output power spectrum compact, the premodulation LPF should have the following properties:

- *Narrow bandwidth and sharp cutoff.* This is needed to suppress high-frequency components.
- *Lower overshoot impulse response.* This condition is required to protect against excessive instantaneous frequency deviation.

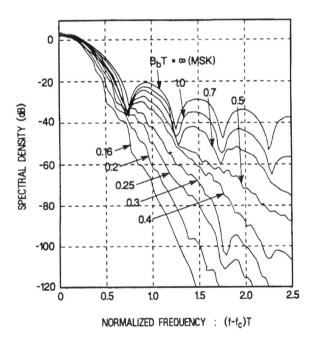

FIGURE 4.9 Power spectra of GMSK. (*Source:* [4]. © 1981 IEEE.)

- *Preservation of the filter output pulse area, which corresponds to a phase shift* $\pi/2$. This condition is required if coherent detection is to be applicable as simple MSK.

Generally, the introduction of the premodulation lowpass filter (LPF) violates the minimum frequency spacing constraint and the fixed-phase constraint of MSK. However, the above two constraints are not intrinsic requirements for effective coherent binary FM with a modulation index of 0.5. Such a premodulation-filtered MSK signal can be detected coherently because its pattern-averaged phase transition trajectory does not deviate from that of the simple MSK.

A Gaussian LPF satisfies the above described characteristics. Figure 4.9 shows the power spectrum of the GMSK signal versus the normalized frequency difference from the carrier center frequency $(f - f_c)T$ with normalized 3-dB down bandwidth of the Gaussian filter B_bT as a parameter. There is a choice between the premodulation filter bandwidth B_b and the bit period T. If $B_b > 1/T$, the waveform is essentially MSK. If $B_b < 1/T$, since the change of symbol cannot reach its next position in the time allotted, the constellation becomes blurred due to the remnants of the previous symbol, as shown in Figure 4.7(c). A B_bT product of 0.5 used in CDPD introduces little intersymbol interference but provides considerably better adjacent channel performance.

The instantaneous frequency variations in GMSK and the corresponding eye diagrams are shown in Figure 4.10.

4.6 MAC LAYER FUNCTIONS

From the point of view of the open system interconnection (OSI) model, the data link layer protocol is divided to two sublayers:

- The lower sublayer of the data link layer resides above the physical layer and is called the media access control (MAC) layer;
- The upper sublayer constitutes the physical layer–independent portion of the data link layer and is called the mobile data link protocol (MDLP).

This section covers the detail functionality of the MAC layer that resides above the physical layer [5,6]. The MAC layer contains the core procedures that allow efficient error-free transmission of information over the air.

4.6.1 Main MAC Layer Functions

The main function of the MAC layer is to convey information between the logical link layers of the MDBS and M-ES across the airlink interface. The frames from the link layer are further block encoded to contain forward error-correction codes. Addi-

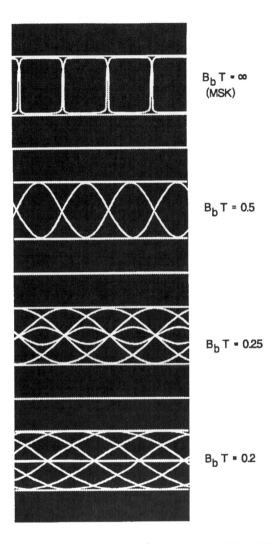

FIGURE 4.10 Instantaneous frequency variation of GMSK. (*Source:* [4]. © 1981 IEEE.)

tional frame synchronization information is added and transmitted on the forward (MDBS to M-ES) and reverse (M-ES to MDBS) channel streams.

The forward and reverse directional channel streams differ in their operational characteristics as illustrated in Figure 4.11:

- The forward channel is a contentionless broadcast channel carrying continuous information from the MDBS to the M-ESs. Data interruption on the for-

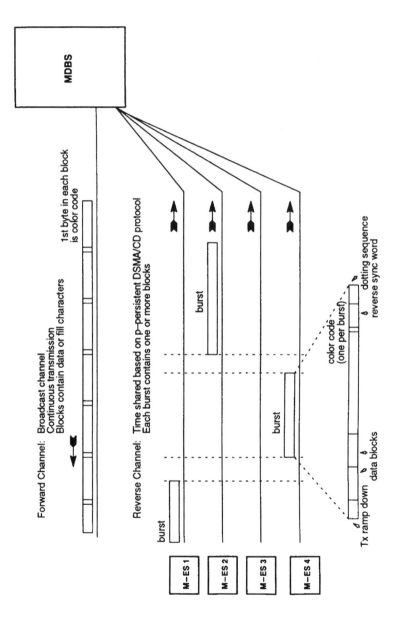

FIGURE 4.11   Forward and reverse channel transmissions.

ward channel stream will occur temporarily when the MDBS switches the RF channel that is being used on the channel stream.

- The reverse channel is a shared channel that can be used by all the M-ESs within the range of the cell boundary covered by the forward channel stream. Multiple M-ESs have to compete for using the reverse channel and transmit data in *bursts* when a channel is acquired. If more than one M-ES transmits simultaneously, resulting collisions will generate blocks with errors.

Uncorrectable block errors due to collision or noise in the radio medium in the reverse channel can be recovered by retransmissions at the MAC. The block errors in the forward channel due to noise cannot be recovered at the MAC level. These will reflect as frame errors in the link level and will have to be recovered by retransmissions at the link layer (i.e., the MAC implementations in the M-ES will have an additional functionality related to accessing reverse radio channels and recovery procedures for retransmission of blocks).

4.6.1.1 Data Transformation

The MAC layer receives information for transmission from the link layer as frames and forwards them to the physical layer in sequence of bits. The transformation is carried out in reverse when receiving date from the physical layer. The different steps of the transformation from frames to bits are detailed below.

Frames to Bit Stream

The initial task of the MAC layer is to convert the sequence of frames it receives from the link layer to a bit-stream in preparation for breaking the bit-stream into blocks. The specific actions that need to be carried out are as follows:

- Individual frames need to be checked for validity. All frames have to be an integral number of octets. The size of the frame shall not be less than 16 bits (2 octets) and shall not be greater than 1,088 bits (136 bytes). Invalid frames will be simply discarded.
- The frames are delimited by the flag sequence (0x7e), an opening flag, and a closing flag. In a sequence of frames, the closing flag of a preceding frame can serve as the opening flag of a succeeding frame. In the reverse channel, the opening flag of the first frame in a burst may be omitted. The last frame in a burst should always have a closing flag.
- Bit-stuffing will be carried out to within the data portion to remove the occurrence of flags or abort sequences (more than 6 consecutive 1's). That is, a 0 bit will be inserted after all sequences of five consecutive 1's.
- Interframe filling may use flags or a marking condition, which can be selected during configuration. The MAC will be required to use interframe fills when

there are no frames to transmit in the forward channel or to fill the last block in a reverse channel burst.

A marking condition lasting less than six bit times will be interpreted at the receiver as an invalid short frame and will be discarded.

Bit Stream to Blocks

The basic unit of transmission used by the MAC layer is a fixed-length error control block that is comprised of the following information 282 bits (47 six-bit symbols) of data extracted from the bit-stream created from the frames and 96 bits (16 six-bit symbols) of Reed-Solomon (RS) forward error-correcting code, making up a nominal block length of 378 bits (or 63 six-bit RS symbols).

After generating the RS codes, the information and the parity bits are exclusive-ORed with a pseudorandom binary code. This process is intended to eliminate the long string of 1's or 0's that may be present in a block after RS coding.

Reed-Solomon Codes Addition of RS codes during the construction of the RS blocks introduces a coding overhead of nearly 33%. The RS codes fall into a general category of burst error-correcting codes and are capable of providing error-correction capability of up to 8 symbols. However, the number of correctable symbols is dependent on the algorithm implemented and most implemented systems correct 6 or 7 symbol errors in a block.

Error-correcting codes fall into two major categories—block codes and convolutional codes [7]. In block codes, the transmitted information, the *code words*, are always of fixed length. These codes are characterized by (n, k) parameters where n represents the total number of bits in a code word and k represents the information bits. Convolutional codes are used in cases where the information is a contiguous string of bits of undefined length. It derives its name from the fact that the encoding involves convolution of the input sequence(s) with the impulse response of the encoder(s).

Hamming codes are a simple subset of linear block codes where a generator matrix and an associated parity matrix can be used to generate a general class of single error-correcting codes of the form $(2^m-1, 2^m-m-1)$. The technique for building Hamming codes breaks down for multiple error correction.

Cyclic codes form another important subset of linear block codes. Whereas Hamming codes are described by means of simultaneous linear equations, the cyclic codes are described by polynomials with coefficients from Galois field GF(q). Code words are represented as polynomials $V(x)$, which are multiples of a single polynomial $g(x)$ called the generator polynomial of the code. Cyclic codes provide a general structure that can be extended to generate multiple error correcting codes (BCH codes).

The desire for error-correcting codes with higher and higher error-correction capabilities lead us out of the realm of random independent error correction into that of random burst error correction. It is seen that t error-correcting codes with large t have low rates and are therefore inefficient. Whereas BCH codes can be defined over the general Galois field $GF(q^m)$, the important subclass, the RS codes result from the special case $m = 1$ and so are constructed from symbols of the Ground field $GF(q)$. The t error-correcting codes over $GF(q)$ can be converted into a burst error-correcting code over $GF(p)$ where $q = p^m$ by interpreting the symbols of $GF(q)$ as m-tuples of symbols of $GF(p)$. In the binary case, an (n, k) RS code with symbols from $GF(2^m)$ has the following parameters:

$t =$	number of correctable error symbols
$n = 2^m - 1$	code length in symbols
$k = n - 2t$	number of information symbols
$n - k = 2t$	number of parity symbols
$d_0 = 2t + 1 = d_{min}$	designed distance = minimum distance

The relevant parameters for the RS-code that is used in the CDPD MAC layer are as follows:

$m = 6$	six bits per symbol
$t = 8$	correctable error symbols
$n = 2^6 - 1 = 63$	63 symbols in code word
$k = 63 - 16 = 47$	47 information symbols
$n - k = 16$	16 parity symbols

For encoding purposes the bits in the block are considered to form a sequence of 63 6-bit RS code symbols $\{s_i\}$. These symbols are transmitted in sequential order from s_{62} to s_0. The least significant bit of each code symbol is transmitted first. The information bits is contained in code symbols s_{62} to s_{16} and the RS parity bits in code symbols s_{15} to s_0.

The generator polynomial is defined as, $g(x) = (x + a^1) \cdot (x + a^2) \cdot \ldots (x + a^{16})$ where a is the primitive field element [010000] under the irreducible polynomial $x^6 + x + 1$. Then the RS parity symbols are the coefficients of the remainder polynomial $r(x) = s_{15}x^{15} + \ldots + s_1x^1 + s_0$ formed as a result of dividing the shifted message polynomial $x^{16}m(x) = s_{62}x^{62} + \ldots + s_{16}x^{16}$ by $g(x)$. Here, the message polynomial $m(x)$ contains the information bits in code symbols s_{62} to s_{16} (47 information symbols) represented as, $m(x) = s_{62}x^{46} + \ldots + s_{16}x^0$.

Pseudorandom Number Coverage When an RS block is generated, the Information field may contain long strings of 1's and 0's. Certain types of modulators and demodulators may have difficulty in accurately tracking the long continuous string of 1's or 0's. To increase the likelihood of bit transitions in the block during transmissions, the 378-bit block is exclusive-ORed with a pseudorandom binary sequence. Each block, therefore, has to be again exclusive-ORed with the same sequence after reception and before RS decoding.

Pseudorandom binary sequences generated by the shift register techniques by using the underlying Galois mathematical structure have a close to randomness property (properties similar to that of white noise) and are called maximum length shift register (MLSR) sequence or m sequences. These sequences are characterized by the fact that their autocorrelation function peaks at $t = 0$ and has a flat and low value for all other values of t. The properties exhibited by these PN sequences are as follows:

- Relative frequencies of 0's and 1's are equal.
- Run lengths of 0's and 1's are similar to those in a Bernoulli trial: half of all run lengths are unity; one quarter are of length 2, and a fraction $1/2n$ of all runs are of length n for all valid n.
- The autocorrelation property; if the random sequence is shifted by any number of bit positions, the resulting sequence will have an equal number of agreements and disagreements with the unshifted sequence. These properties have been found to be useful in eliminating the continuous 1 or 0 pattern in a stream when it is exclusive-ORed with a PN-sequence.

The PN sequence used in by the MAC layer is generated using the generator polynomial

$$g(D) = D^9 + D^8 + D^5 + D^4 + 1$$

and this is initialized with a binary value 111000101 every 378 bits. It has to be noted, however, that the periodicity of this sequence is $2^9 - 1 = 511$ bits and only the leading 378 bits are used in the PN coverage.

Blocks to Bursts

For the forward channel, the channel stream is continuous and therefore blocks will be continuously assembled and transmitted on the forward channel stream. During periods when there are no forward frames to transmit, the MAC layer will fill the transmissions with either flags or marking bits (1's). Further, a synchronization pattern, decode flags, and channel status flags are interleaved within the data.

Reverse channel transmissions will be in bursts that have a dotting sequence and a reverse synchronization pattern prefixed to the RS blocks. Further, continuity indicator bits will be interleaved within the RS block. For a full-duplex M-ES, there will be a configuration-defined maximum number of blocks that can be sent in a single burst, whereas a half-duplex M-ES can transmit only one block in a reverse burst.

4.6.1.2 Medium Access Management

The channel access protocol is defined as a slotted nonpersistent digital sense multiple access protocol with collision detection (DSMA/CD). Slotted implies synchronism in that the reverse channel transmission processing in all M-ESs using the same forward channel will be synchronized. For this purpose a microslot is defined as consisting of 10 RS symbols, and a microslot is equivalent to a transmit time of 60 bits, which is 3.125 ms. Microslot clocks in the M-ESs will be in lockstep to the microslot boundaries in the received forward channel. Nonpersistent algorithm means that only those messages (frames) that arrive when the channel is sensed idle are transmitted. All other frames are immediately rescheduled as if a collision had occurred. This technique is different from both 1-persistent DSMA and *p*-persistent DSMA, both of which require transmission of messages quicker with respect to their time of arrival than that for the nonpersistent case.

The forward channel contains two flags that the MAC layer in the M-ESs use to contend and acquire the reverse channel for transmitting data to the MDBS.

- *Channel status flag.* The channel status flag signifies whether the reverse channel is busy or idle. The MAC layer of an M-ES that has data to transfer defers the transmission until the channel status flag indicates that the channel is idle. This helps to avoid collisions in the reverse channel.
- *Decode status flag.* This indicates to the M-ES that has obtained access to the reverse channel whether the block that was transmitted has been decoded without any errors. If there has been a collision, the two (or more) transmitting M-ESs will discover that they have collided by looking at this flag. The MDBS that generates this bit after decoding each block will not be able to discover whether the error was due to collision or due to noise in the channel. Further, even if the collisions were such that the reverse bursts were not even synchronized, the MDBS is supposed to be sending a default value of "decode fail," and the transmitting M-ESs would correctly interpret the collision condition.

The MAC layer, further, uses the following configurable parameters in its management of access to the channel. Time intervals in the descriptions refer to number of microslots (3.125-ms units).

- *Maximum blocks in a burst.* This parameter limits the length of time an M-ES can hold the reverse channel once it has acquired it, thereby allowing other M-ESs to share the reverse channel.
- *Minimum idle time.* Once an M-ES has used the reverse channel for its transmission burst, it cannot attempt to transmit another burst for this amount of time. This parameter again helps to allow the other M-ESs to be successful in acquiring the channel.

- *Maximum/minimum backoff interval.* When an M-ES MAC discovers that the reverse channel is busy, it has to backoff for a random time interval before checking for the channel to be idle. The initial backoff time will be a uniformly distributed random number between 0 and 2^{min}; the next backoff time will be a uniformly distributed random number between 0 and $2^{(min+1)}$, and after every attempt it is recomputed until the nth attempt where the equivalent backoff interval is between 0 and 2^{max}.

4.6.1.3 Channel Stream Synchronization

The timing reference for the slotted DSMA protocol is established by the synchronization words transmitted in the forward channel. All M-ESs when acquiring the RF channel synchronize to the master microslot clock derived from the forward channel. The forward synchronization word further enables the M-ES to establish RS block boundaries and to locate the channel status and decode status flags. Transmission by the M-ES may begin only at the beginning of microslot boundaries.

Similarly, the reverse channel bursts contain synchronization words that support the reception of the bursts and decoding the burst that may contain multiple blocks.

Both forward and reverse transmissions contain color codes that allows the CDPD system to detect co-channel interference. A part of the color code enables the M-ES to identify when an area change (channel stream of a different MD-IS) has occurred so that it can carry out procedures required to transfer from one MDIS to another MDIS.

4.6.2 Forward Channel

This section details the format of the channel transmissions in the forward channel and the procedures followed by the M-ES and MDBS.

4.6.2.1 Forward Channel Format

The forward channel transmission structure is shown in Figure 4.10. Each RS-block in the forward channel contains the following elements:

- Reed-Solomon FEC block, which is made up of 47 RS symbols of data and 16 RS symbols of parity bits. These have been covered by the PN sequence.
- The first 8 bits of the 47 data symbols represent the channel color code.
- Interleaved with the above sequence of bits are 7 additional RS symbols containing the following:
  - Forward synchronization word;
  - Decode status flag;
  - Busy/idle status flag.

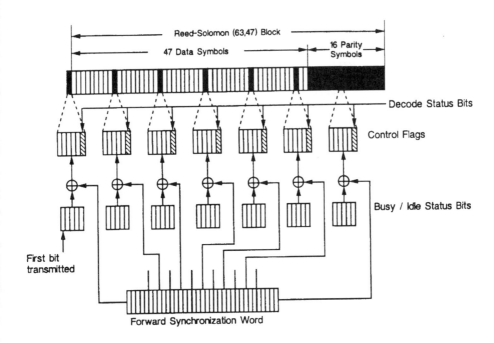

FIGURE 4.12 Forward channel transmission structure. (*Source:* [1].)

Figure 4.12 further shows the interleaving scheme as to how the RS symbols are grouped into microslots. Each microslot starts with a control flag symbol and has 9 additional data symbols forming a 60-bit microslot. Seven of these microslots form a forward channel RS clock. In summary, the sizes and corresponding transmit times of the different components in the forward channel transmission are shown in Table 4.2.

TABLE 4.2
Components in Forward Channel Transmission

Entity	Size	Transmit Time
Microslot	10 symbols; 60 bits	3.125 ms
Block	7 microslots; 420 bits	21.875 ms
Color code	8 bits	
Data (including color code)	47 symbols; 282 bits	
RS parity bits	16 symbols; 96 bits	
Control flag	7 symbols; 42 bits	

Forward Channel Framing and Block Structure

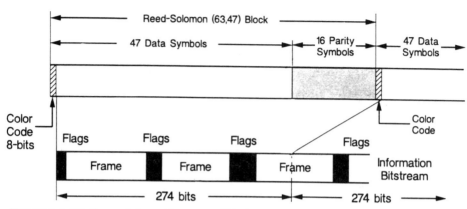

FIGURE 4.13   Forward channel framing and block structure. (*Source:* [1].)

We can observe that each block in the forward direction can transport $(282 - 8)/8 = 34.25$ octets of information.

The mapping of frames to blocks in the forward channel is shown in Figure 4.13. Depending on the size of the frames, both the following cases are possible:

- A block may contain more than one frame if the frame sizes are smaller than 35 bytes.
- More than one block may be used to carry a frame if the frame size is more than 35 bytes.

Synchronization Word

The forward synchronization word is a 35-bit sequence that allows the M-ES MAC layer to identify the FEC block boundaries and to synchronize its timing for the reverse channel microslot clock. The sequence, in the order of bit transmission, is

$$11101\ 00001\ 11000\ 00100\ 11001\ 01010\ 01111$$

Each of the above groups (seven groups) of five bits undergoes the following transformation:

- Exclusive-ORed with the five-bit busy/idle flag.
- A 1-bit decode status flag is added to make each of the 5 bits of the sync word into a 6-bit RS symbol.

The resulting 7 RS symbols are then inserted at equal spacing in the 63-symbol RS block as shown in Figure 4.12.

The specification recommends that the probability of missed sync detection in AWGN should be no worse than using binary correlation with threshold detection, which tolerates up to 5 bits in error. This corresponds to missed sync detection probability of $1.265 \cdot 10^{-6}$ and false sync detection of $1.118 \cdot 10^{-5}$ at 1% BER in AWGN.

Busy/Idle Flag

The busy/idle flag is set to busy when the MDBS detects the presence of 19.2 Kbps GMSK-modulated transmission on the reverse channel. A 5-bit sequence with all 0's indicates a channel busy condition and all 1's indicates a channel idle condition. A majority decoding logic is recommended for use in the M-ES.

Decode Status Flag

The decode status flag is also a 5-bit sequence indicating whether the MDBS MAC was able to decode the preceding block received on the reverse channel. All 0's indicates decode success and all 1's indicates decode failure.

The bits of the decode status flag are transmitted in sequence one bit per microslot as part of the control flag, as shown in Figure 4.12. The first bit of this flag will be transmitted in the third control flag following the complete reception of the reverse channel block to which the decode status applies. Although the M-ES retrieves only 5 bits of the decode status flag, the MDBS normally transmits 6 or 7 bits depending on the exact timing between the forward and reverse channels. The default value of the decode status bit is decode failure.

4.6.2.2 Forward Channel Processing

Each M-ES tuned to the MDBS forward channel transmit frequency receives the same forward channel data stream. MDBS will be continuously transmitting in the forward channel while an RF channel is available for CDPD. When there is a requirement to switch RF channels, the old carrier is simply turned off and the new RF channel, when available, is tuned and the forward transmission continues. The forward channel stream does not carry any MAC-level information related to the RF channel switch. In special cases, higher level messages are sent on the forward channel stream informing the M-ESs that an RF channel switch is to occur.

The MDBS inserts 8-bit color code at the beginning of every block made up of 274 data bits, and the resultant 282 bits are then encoded as a RS code word. The synchronization word and MAC control data, such as busy/idle flag and decode status bits, are combined with RS blocks prior to transmission.

Busy/Idle Flag

The MDBS uses the following states and the following state transitions in handling a normal burst from an M-ES.

Channel Idle State

MDBS enters this state after the initial acquisition of the channel stream and will set the flag to idle to indicate that there are no reverse channel transmissions.

SYNC Search State

When the MDBS detects GMSK-modulated data on the reverse channel, it sets the idle bit to busy and enters the SYNC search state looking for a SYNC pattern.

SYNC Lock State

On detection of the SYNC pattern, the MDBS switches to this state receiving RS blocks and keeps the busy/idle status busy. While examining the EOT indicator flag in the reverse channel, when it becomes set it transitions to the busy/hang state while still keeping the busy flag set.

BUSY Hang State

The MDBS keeps the busy flag set to busy until the transmitting M-ES has been given the 2-ms ramp downtime to switch off its transmitter. This works out to be equivalent to 33 bit times after the reception of the last bit of the block that contained the end of transmission bit set to 1. After this delay, the MDBS MAC transitions to the channel idle state.

SYNC Hold State

SYNC Hold state is a transient error-recovery state that the MDBS can transition from the SYNC lock state when the channel degradation results in temporary loss of data.

4.6.3 Reverse Channel

This section details the format of the channel transmissions in the reverse channel and the procedures followed by the M-ES and MDBS.

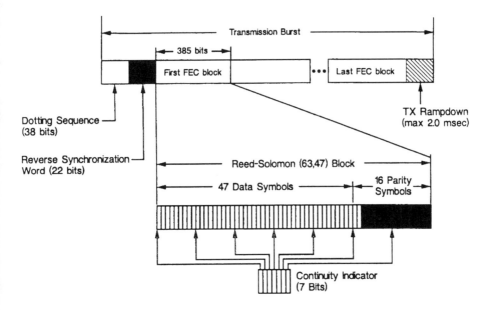

FIGURE 4.14 Reverse channel transmission structure. (*Source:* [1].)

4.6.3.1 Reverse Channel Format

Whereas the forward channel carries a continuous stream of data, the reverse channel contains bursts from multiple M-ESs. The overall structure of a burst in the reverse channel is shown in Figure 4.14. A burst contains the following elements:

- A 38-bit dotting sequence of 101010... pattern coinciding with the ramp-up time of the reverse channel transmitter in an M-ES. This is a preamble and allows an MDBS to detect a burst (energy) and also to recover bit timing.
- A 22-bit reverse channel synchronization word.
- One or more Reed-Solomon blocks.
- A 2-ms ramp-down time while transmitting a 0 to complete the burst.

In summary, the sizes and corresponding transmit times of the different components in a reverse channel burst are shown in Table 4.3.

TABLE 4.3
Components in Reverse Channel Transmission

Entity	Size	Transmit Time
Dotting sequence	38 bits	1.979 ms

TABLE 4.3 (continued)

Entity	Size	Transmit Time
Sync word	22 bits	1.146 ms
First block color code	8 bits	
First block data	46 symbols; 276 bits	
Subsequent block data	47 symbols; 282 bits	
RS parity bits	16 symbols; 96 bits	
Continuity indicator (interleaved)	7 bits	
Block	385 bits	20.052 ms

Each block of data transmitted in the reverse channel contains the following:

- The first 8 bits of the Information field of the first block in a burst contains the channel color code.
- All blocks other than the first block contain 47 6-bit RS symbols of data and 16 6-bit symbols of parity information. These bits have been covered by the PN sequence.
- Interleaved with the above sequence of bits are 7 continuity indicator bits.

This is illustrated in Figure 4.15.

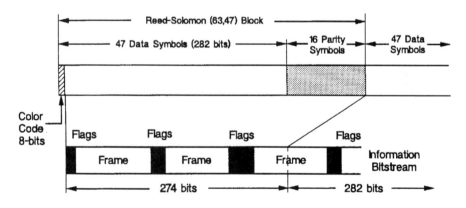

FIGURE 4.15 Reverse channel framing and block structure. (*Source:* [1].)

4.6.3.2 Reverse Channel Processing Overview

MDBS Procedures

The MDBS has continuous access to the reverse channel stream. The MDBS notifies to listening M-ESs the presence of data on the reverse channel and the decode status of the incoming data blocks from the reverse channel using the busy/idle flag and the decode status flag in the forward channel stream.

The implementation of the detection of data on the reverse channel must be such that M-ES transmissions within 8 bit times of the last bit of an idle flag should result in the next busy/idle flag to be set. This is a key requirement for the efficient and correct operation of the DSMA/CD protocol used for channel access by the M-ESs. MDBS, after receiving the sync word, decodes the data blocks within two microslots and sets the next decode status flag with the result. Once the flag is set to indicate decode success, it remains in that state until a decode failure occurs or the last block in a burst is successfully received and has been signaled as a success.

MDBS determines the end of burst if one of the following conditions occurs:

- Normal termination indicated by the continuity indicator.
- Detection of an undecodable block. Subsequent blocks in the burst are discarded until the EOT continuity flag is received or absence of data on the reverse stream is detected.
- Failure to detect the presence of data after the last continuity indicator.
- Received blocks in the burst exceed the configured maximum.

More details on timing constraints are given in the next section.

M-ES Procedures

The efficient and specification-compliant implementation of the DSMA/CD protocol for the reverse channel access is key to proper MAC operation. The M-ES typically uses the following internal states to comply with the access procedures.

IDLE State The state the M-ES stays in when there are no frames to transmit. When it reenters this state after transmitting a burst, it will have to remain for minimum idle time.

DEFER State When the reverse channel is busy, and the M-ES performs a random backoff, it stays in this state.

BACKOFF State The M-ES layer entity will be in this state when attempting to re-transmit one or more blocks that have been indicated as decode failure by the MDBS. The first block to be retransmitted will be the block containing the beginning of the frame that is contained in the first block that was marked as decode failure.

TRANSMIT State When the M-ES is in this state, it transmits a bit stream containing interleaved continuity indicators. The M-ES can enter this state from any one of the IDLE, DEFER, or BACKOFF states.

DECODE WAIT State This is a temporary state in which the M-ES waits a brief amount of time to determine whether the last block of its last transmission was successfully received by the MDBS or not. The time allocated for this is 7 microslots and consists of the following:

- Two microslots or 120 bit times (6.25 ms) to decode the RS block.
- Five microslots or 300 bit times (15.625 ms) to indicate the decoding status to the M-ES using the decode status flag.

For a detailed specification of the recommended procedures, the reader should refer to the CDPD specifications.

4.6.4 Channel Timing Requirements

This section outlines some of the timing requirements that are critical for proper operation of the airlink between the M-ES and the MDBS. Strict adherence to the specifications is essential for successful interoperation between M-ESs and MDBSs manufactured by different vendors.

4.6.4.1 Start of Burst Timing

An M-ES, after performing the MAC level procedures for reverse channel access, is ready to transmit and has to initiate the transmission within 8 bit times (416.7 µs) of the last bit of an idle flag in the forward channel. This is the worst case delay and implementation-dependent actual delay is expected to be around half this value. This delay accounts for the following:

- Propagation delays in the M-ES receiver circuit (to identify the flag);
- Processing delays in the demodulator;
- M-ES MAC protocol processing;
- Propagation delays in the M-ES transmitter circuit.

The burst begins with the dotting sequence, and the transmitter ramp-up must complete within 2 ms, as shown in Figure 4.16. Further, the integrated value of the power within the first 38 bit times (dotting sequence transmission—1.979 ms) must be at least 50% of the similar value when taken during steady state transmission. All out-of-band frequency transmissions should completely cease within 3 ms.

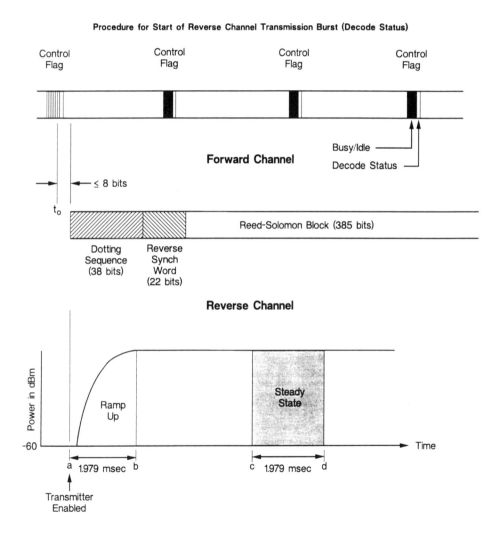

FIGURE 4.16 Start of reverse channel burst timing. (*Source:* [1].)

4.6.4.2   End of Burst Timing

The end of burst timing characteristics of an M-ES has an impact on the efficient sharing of the reverse channel. More specifically, timing requirements must be defined so that the end of burst of an M-ES that is currently using the reverse channel and the start of burst of a new M-ES must be allowed to occur as close together as possible in the time-multiplexed reverse channel, but without the danger of these bursts overlapping one another.

Figure 4.17 shows the timing relationships between the continuity indicator bit in the reverse channel and the busy/idle flag in the forward channel. The key requirement is that the MDBS has to declare the reverse channel as busy until the transmitting M-ES after completing the last bit of the burst has had sufficient time to ramp-down its transmitter (2 ms maximum). This requirement can be broken down further as follows:

- 2 ms is equivalent to 38 transmitted bit times.
- Since the busy/idle flag is 5 bits and the decision at the MDBS MAC is taken when the first bit of this flag is transmitted, the MDBS must maintain busy condition at least for 33 bit times after receiving the last bit of the last reverse block.
- The above timing is required at the M-ES antenna as the reference. However, as seen in Section 4.6.4.1, the M-ES requires about 8 bit times propagation delay and the actual extended busy time in the MDBS MAC is about 25 bit times.

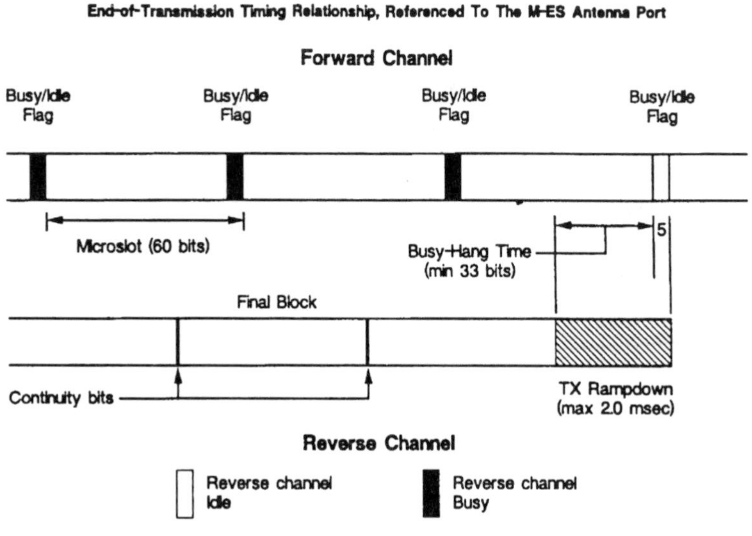

FIGURE 4.17   End of burst timing. (*Source:* [1].)

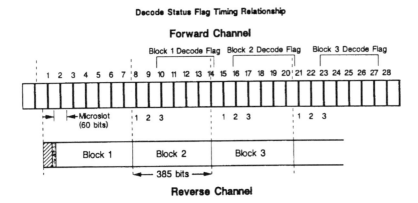

FIGURE 4.18 Decode status flag timing. (*Source:* [1].)

- Propagation delays at the MDBS end can be used to further decrease the busy time to less than 25 bit times.

4.6.4.3 *Timing of Decode Status Flag*

The MDBS sets the decode status flag when the RS decoding of a received block is unsuccessful or the color code received in the reverse burst does not match the transmitted color code. The third control flag following the end of the received block will carry the resultant decode flag, as shown in Figure 4.18. This gives the MDBS a worst- case time of 2 microslot times to decode a RS block.

4.7 SUMMARY

The airlink protocol is a key component of the CDPD protocol architecture and serves to define the robustness of the airlink data transfer capability. Reed-Solomon coding provides significant forward error-correction capability. Further, MAC-level retransmission can be used to recover from uncorrectable block errors in the reverse link. This additional error-recovery procedure at the MAC level, that does not involve higher layers of the protocol stack, provides efficient recovery from errors due to collision and from poorly performing M-ESs transmitting at the weaker end of the power scale.

Compliance with the specification and the resulting interoperability characteristics between components of different vendors of MDBSs and the CDPD modems will have a major impact on the adoption of CDPD as a suitable technology for digital wireless data transfer by potential customers.

References

[1]   CDPD Specifications Version 1.1, CDPD Forum.

[2]   EIA-553: Mobile station-land station compatibility specification.

[3]   Pasupathy, S., "Minimum Shift Keying: A Spectrally Efficient Modulation," *IEEE Communications Magazine*, July 1979, pp. 14–22.

[4]   Murota K., and K. Hirade, "GMSK Modulation for Digital Mobile Radio Telephony," *IEEE Transactions in Communications*, Vol. Com 29, No. 7, July 1981, pp. 1044–1050.

[5]   Tasaka, S., *Performance Analysis of Multiple Access Protocols*, Cambridge, MA: The MIT Press, 1986.

[6]   Proakis, J. G., and M. Salehi, *Communication Systems Engineering*, Englewood Cliffs, NJ: Prentice Hall, 1994.

[7]   Lin, S., and D. J. Costello, Jr., *Error Control Coding: Fundamentals and Applications*, Englewood Cliffs, NJ: Prentice Hall, 1983.

CHAPTER 5

▼▼▼

LINK LAYER PROTOCOL

5.1 INTRODUCTION

The data packets between the M-ES and another end system (this could be another M-ES or a fixed-end system) are processed by the mobile data intermediate system (MD-IS). This transfer makes use of the MDBS as a relay between the MD-IS and the M-ESs.

Communications between the M-ESs and the mobile data intermediate systems occur primarily at the logical link control and subnetwork convergence protocol layers. These protocol layers are part of the collective set of protocols used across the CDPD airlink. The complete airlink protocol profile is as described in Figure 5.1. Mobile data link protocol (MDLP) is a protocol that operates within the data link layer of the OSI layered architecture to provide logical link control services between the M-ESs and the MD-ISs. This chapter will attempt to describe the services provided by the link layer, its model of operation and its interfaces with its lower and upper layers within the context of the overall CDPD protocol architecture. It will also describe some typical flow walkthroughs that comprise the M-ES to MD-IS communications. Note, however, that this chapter is not intended to be a formal definition of the link layer.

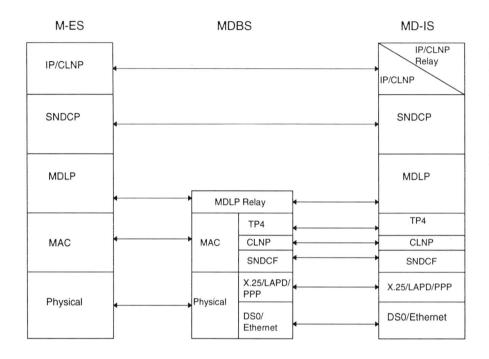

FIGURE 5.1 Airlink protocol profile.

5.2 OVERVIEW OF THE LINK LAYER

The CDPD airlink interface provides access to the CDPD network using either the Internet protocol (IP) defined in [1] or the ISO connectionless network protocol (CLNP) defined in [2]. The network layer protocol data units (NPDUs, or more commonly referred to as packets) are transmitted across a mobile data link between the M-ES and the MD-IS using the MDLP. The MDLP is a modified form of the link access protocol-D (LAPD) defined in [3]. The necessary subnetwork-dependent convergence protocol (SNDCP) required to transform packets, generated by multiple layer 3 protocols into a format appropriate for MDLP will be discussed in Chapter 6.

The transformations undergone by an individual network layer packet for transmission across the airlink are described in detail in Section 4.2 and are summarized as follows:

- The SNDCP layer receives the packet from the network layer and optionally compresses the packet header, then optionally compresses the packet using V.42 bis data compression techniques. It then segments the packet and provides each segment with a segment header. Each segment is then encrypted and forwarded to the MDLP protocol entity.

- At the MDLP layer, each segment is encapsulated within a frame header and a number of functions are performed on this data segment, as described in Section 5.2.1.
- Each of the frames generated by the MDLP layer is transmitted out onto the RF channel using the services provided by the MAC and the physical layers.

5.2.1 Services Offered by This Layer

The MDLP layer has been derived extensively from the LAPD protocol with additional concepts taken from the HDLC protocol defined in [4]. Additionally, some functions needed for mobility-specific operations have been added to the specification of this layer. As a matter of convention, while referring to the link layer, the layer management entity associated with this layer is also included.

The data link layer is responsible for establishing the data link between the M-ES and the MD-IS. The M-ES endpoint of the data link is termed as the *user side* of the data link, whereas the MD-IS is termed as the *network side* of the data link. The purpose of the MDLP is to convey information between the user side and the network side layer 3 (network layer) entities across the airlink interface. The layer 3 entities that use the services of the MDLP include the mobile network registration protocol (MNRP), security management entity (SME), Internet protocol (IP), and the connectionless network layer protocol (CLNP). At the MDBS, the MDLP relay functionality accepts input from the MDBS radio resource management (RRM) for transmission of broadcast management messages to the channel stream.

This layer, while being quite useful in any wireline data communications, is especially useful in wireless data communications, due to the inherent characteristics of the radio transmission medium. These characteristics are typically an error rate that is higher than wireline communications, along with the fact that radio frequencies are a scarce and expensive resource. The most useful service this layer would provide to higher layers would be its error detection and recovery capabilities, thereby attempting to alleviate the error recovery and consequent retransmissions that these higher layers may generate.

The functionality of the MDLP is achieved by a variety of functions that this layer performs. These functions are resident mainly at the MD-IS, but a small set of functions is also defined for the MDBS.

5.2.1.1 MDLP at the MD-IS

The first function is the assignment of temporary equipment identifiers (TEIs), used for addressing individual M-ESs within the serving area for an MD-IS. A channel stream can contain one or more logical data link connections. Discrimination between individual data link connections is by means of the TEI contained in each frame. Negotiation of link parameters is possible on a link establishment.

After the TEI has been assigned, the MDLP layer enables the transferring of link protocol data units (LPDUs) or frames across the mobile data link between the subscriber device and the MD-IS. This transfer is achieved in one of the following ways:

- Unacknowledged point-to-point delivery of frames between the M-ES and the MD-IS.
- Reliable sequenced acknowledged point-to-point delivery of frames between the M-ES and the MD-IS.
- Broadcasting unacknowledged frames from the MD-IS to all M-ESs.

The link layer detects transmission, format, and operational errors on the data link connection and has procedures to recover from such errors. It also provides flow control between the M-ES and the MD-IS in the event that either data link entity becomes busy.

The MDLP makes use of primitives to interact with the local management entity (layer management entity). This interaction can be to inform the management entity of successful establishment of the connection or to inform it of unrecoverable errors.

An important function defined as part of MDLP is that of maintaining a data link connection in sleep mode to allow implementation of M-ES power conservation mechanisms.

The MDLP also provides an interface into the accounting meter functionality at the MD-IS. This allows the MD-IS to collect usage information for the M-ES to MD-IS link.

5.2.1.2 *MDLP at the MDBS*

At the MDBS, the MDLP functionality provides a link function between the channel stream and the MDBS-MD-IS link. The MDLP relay accepts MDLP frames from one side and forwards them to the other according to the defined mapping. The MDLP relay also accepts input from the MDBS RRM entity for transmission to the M-ESs.

5.2.2 Interfaces With Other Protocol Layers

The link layer has interfaces to its upper and lower layers within the context of the overall CDPD protocol architecture. Figure 5.2 illustrates the MDLP reference model with respect to interactions with other protocol layers and to the local layer management entity [5,6].

These interfaces are explained in the following sections.

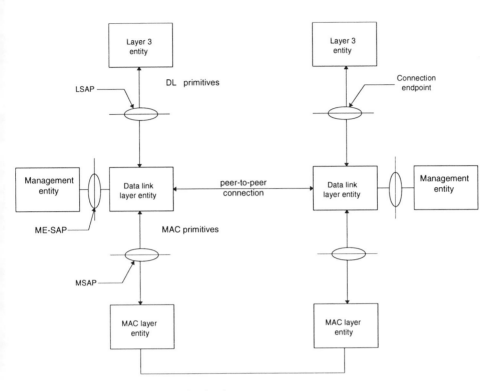

FIGURE 5.2 MDLP interaction with other layers.

5.2.2.1 *Interface With MAC*

The MDLP to MAC interface consists of three parts:

- Transmitter interface;
- Receiver interface;
- Control interface.

The primitives used for the Transmitter Interface are as follows:

- MAC_DATA.request to send a MDLP frame down to the MAC layer. This event is posted by the MDLP layer to indicate to the MAC layer that a frame has been posted to the transmit queues at the MAC layer.
- MAC_DATA.confirm from MAC to MDLP to indicate that a frame that was received at the MAC from the MDLP layer has been transmitted out onto the channel stream.

The primitive used for the receiver interface is a MAC_DATA indication from that MAC layer to indicate that a frame has been received from the channel stream.

There are no standard primitives defined for the control interface. These would typically be for management control functions like the handling of a "ZAP" command. The ZAP command is used by the MD-IS to request the M-ES data link layer entity to stop transmission of any kind of MDLP frames at the user side MDLP entity. In this case, the MDLP may want MAC to halt sending data at the end of the current frame. Likewise, the MDLP may indicate to the MAC that it may now resume sending data to the network side MDLP entity.

5.2.2.2   Interface With Layer 3

The MDLP to SNDCP interface consists of three parts:

- Transmitter interface;
- Receiver interface;
- Control interface for layer 3 management.

The primitives used for the transmitter interface are as follows:

- DL_DATA.request posted by the SNDCP layer to the MDLP. This indicates that the SNDCP has a segment that needs to be sent out on a data link, using the acknowledged class of service.
- DL_UNITDATA.request used by the layer 3 to request that a message unit be sent using the procedures for unacknowledged information transfer service.

The primitives used for the receiver interface are as follows:

- DL_DATA.indication posted by the MDLP to the layer 3. This indicates that at least one frame was received by means of the acknowledged information transfer service.
- DL_UNITDATA.indication posted by the MDLP to the layer 3. This indicates that layer 3 packets have been received by the link layer using the unacknowledged information transfer service.

The primitives used for the control interface are as follows:

- DL_ESTABLISH.request posted by the layer 3 to the MDLP to request establishment of multiple frame operation. Multiple frame operation is used for the acknowledged information transfer service.
- DL_ESTABLISH.indication posted by the MDLP to layer 3, indicating that the link layer has entered the multiple frame operation for the TEI indicated.

- DL_ESTABLISH.confirm used by the MDLP to inform layer 3 that the activity towards establishment of the multiple frame operation has been completed.
- DL_RELEASE.request, DL_RELEASE.confirm and DL_RELEASE.indication towards the procedures for terminating a previously established multiple frame operation for a particular TEI or for reporting an unsuccessful establishment attempt.

5.2.2.3 *Interface With the Local Layer Management Entity*

Only unacknowledged information transfer service is provided to the local layer management entity so that the two peer data link layer management entities (on the M-ES and the MD-IS) may communicate.

- MDL_UNITDATA.request is used by the layer management to request that a message unit be sent using the procedure for unacknowledged information transfer service.
- MDL_UNITDATA.indication posted by the MDLP to the layer management to indicate the arrival of a message unit destined for the local layer management entity. This message unit is received by means of the unacknowledged information transfer service.

Apart from the information transfer between the peer layer management entities, administrative services are also provided by the link layer. The service primitives used for these administrative services are for the assignment and removal of TEI values, data link connection parameter passing, sleep mode supervision service, intra-area mobility service, the zap service, and the loopback test service.

5.3 MODEL OF OPERATION

Two types of operations of the link layer are available for layer 3 information transfer across the CDPD airlink. These are the unacknowledged and acknowledged modes.

5.3.1 Unacknowledged Operation

The layer 3 information is sent out in unnumbered information (UI) frames, in this type of model. These frames are not acknowledged at the data link layer. There is no error recovery mechanism or flow control defined on this type of operation. UI frames can be sent to a specific address or TEI, or broadcast to multiple M-ES addresses.

5.3.2 Acknowledged Operation

The layer 3 information is sent out in numbered information (I) frames. These frames are acknowledged at the data link layer, with sequence integrity of data link protocol data units being maintained. Acknowledged frames can be sent only to specific end-points or TEIs. This leads to a reliable, sequenced, point-to-point data link connection between the two endpoints. One form of acknowledged information transfer is defined—multiple frame operation (MFO). This model includes error detection and recovery mechanisms and flow control procedures.

Multiple Frame Operation

The acknowledged information transfer service works over an established data link connection between the two connection endpoints identified by the assigned TEI. This mode of operation is called the MFO. Layer 3 protocol data units go across this connection in a sequenced manner. Loss of sequence is notified to the peer entity. Flow control is exercised by either data link layer entity.

5.3.3 TEI

The TEI is used by the link layer to address individual subscriber units or M-ESs. This identifier may be one of the point-to-point ones or a broadcast TEI. The point-to-point TEI is assigned by the MD-IS data link layer entity. Broadcast TEI values are predefined and reserved.

All link layer frames contain a TEI that identifies the destination of that frame. The receiving data link layer entity receives frames with the point-to-point TEI if it is aware of that TEI. The receiving data link layer entity analyzes the TEI value in each valid received MDLP frame to determine if the frame is intended for itself. The receiving data link layer entity will also receive all frames with the broadcast TEI values if it has been configured to do so.

5.3.3.1 TEI for Broadcast Data Link Connection

The TEI for a broadcast data link connection is associated with all user side data link layer entities.

A TEI value of zero is reserved for layer 2 management procedures. This TEI value would be used for procedures like TEI Request, TEI Assignment, TEI Check, TEI Check Response, and TEI Removal. It is also used for radio resource management (RRM) forward channel broadcast messages and TEI Notifications used for sleep mode management. PDUs belonging to these management procedures would have the Address field set to 0.

A TEI value of 1 is used by the network side data link layer entity (MD-IS resident MDLP) to transmit layer 3 broadcast NPDUs. This would typically be used for point-to-multipoint multicast or broadcast information transfer.

5.3.3.2 TEI for Point-to-Point Data Link Connection

Point-to-point data transfer uses the defined range of values for the point-to-point data link connections. An M-ES may contain only one TEI used for point-to-point transfer. TEI values in the range of 16 through (227 – 1) are used for point-to-point data link connections. These would provide information transfer services to layer 3 entities.

5.3.3.3 TEI Assignment Procedures

Figure 5.3 shows the sequence of messages leading up to a TEI assignment for a requesting M-ES. The user side management entity transmits the *identity request* message to its peer on the serving MD-IS to request a TEI and negotiate data link layer parameters. On the MD-IS, the network side data link layer management entity generates the *identity assign* message according to defined procedures. If the incoming

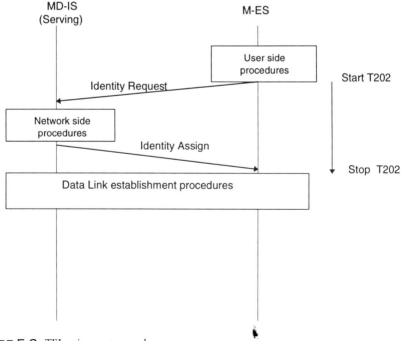

FIGURE 5.3 TEI assignment procedure.

identity request message contains an equipment identity (EID) that is currently as-signed a TEI, the serving MD-IS removes the current TEI value and tears the data link connection associated with that TEI. It then assigns a new TEI value and transmits the corresponding identity request message to the requesting M-ES. As part of these procedures, a TEI check procedure could also be initiated. The TEI check procedures are typically executed if available TEI information or resources are exhausted or if a possible multiple TEI assignment is suspected.

5.3.4 Data Link States

A point-to-point data link layer entity may be in one of three basic states, as shown in Figure 5.4. These states are as follows.

- *TEI-unassigned state*. In this state, a TEI has not been assigned. No NPDU transfer across the airlink is possible. No connection exists while the data link is in this state.
- *TEI-assigned state*. A TEI has been assigned as a result of the TEI assignment procedure. Unacknowledged information transfer is possible. A data link con-nection exists but is not in the multiple-frame-established state.
- *Multiple-frame-established state*. Acknowledged and unacknowledged infor-mation transfer are possible in this state, between the network and a desig-nated user entity. The user entity at the M-ES is addressed by its assigned TEI.

A broadcast data link layer entity is always in the TEI-assigned state. It is capa-ble of only achieving unacknowledged NPDU transfer. Broadcast TEI values are known by prior definitions (as explained in Section 5.3.3.2).

Because MDLP is an asymmetric protocol, the behavior and procedures for the user side CDPD devices (M-ESs) differ from those used by the network side CDPD devices (MD-IS).

5.4   FORMATS OF FIELDS

All link layer exchanges are in frames or LPDUs. All frames contain an integral number of octets, each octet containing 8 bits. The order of bit transmission across the channel stream is defined by the MAC layer. The frame format is defined in Figure 5.5.

Each MDLP frame contains an Address field (defined by the TEI assignment procedures outlined in Section 5.3.3) and a Control field (described in Section 5.4.1). It may also contain an Information field. The Information field, if present, will con-tain the SNDCP header as the first octet(s).

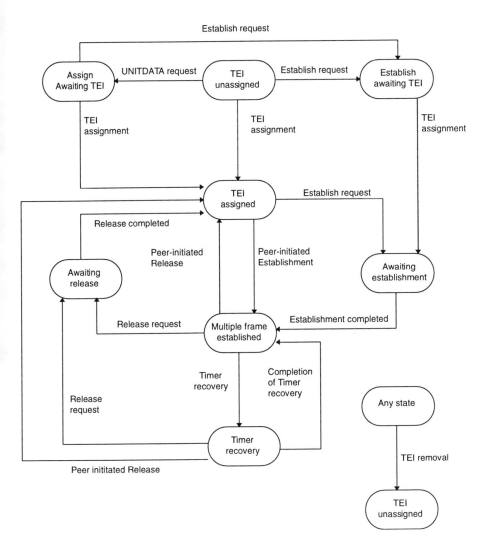

FIGURE 5.4 Data link states.

5.4.1 Control Field Parameters

The Control field identifies the type of frame. It also contains sequence numbers for certain types of Control field formats.

The numbered information transfer format is abbreviated as the *I format*. This is used for information transfer between peer layer 3 entities. Each I frame has an $N(S)$ (transmitter send sequence number) and an $N(R)$ (transmitter receive sequence number) and a P bit (Poll bit) that may be set to 0 or 1.

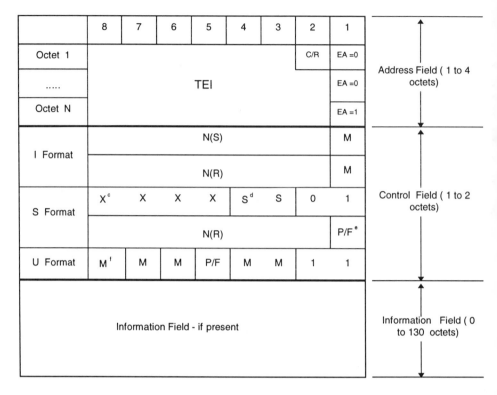

FIGURE 5.5   Frame format.

The supervisory format is abbreviated as the *S format*. This type of frame performs data link supervisory control functions such as acknowledge I frames, request retransmissions of I frames, and request a temporary suspension of I frames. This frame contains the $N(R)$ sequence number.

The unnumbered information transfer and control format is abbreviated as the *U format*. This type of frame is used to perform additional data link control functions and unnumbered information transfers for unacknowledged information transfer. This format does not contain sequence numbers.

5.4.1.1   Variables and Sequence Numbers ($V(S)$, $V(A)$, $V(R)$, $N(S)$, $N(R)$)

The $V(S)$ denotes the sequence number of the next I frame to be transmitted. The $V(S)$ may take on the value 0 through 127. The $V(S)$ gets incremented by 1 with each transmission of an I frame and does not exceed $V(A)$ by more than the maximum number of outstanding I frames.

The $V(A)$ identifies the last frame that has been acknowledged by its peer. The value of $V(A)$ is updated by the valid $N(R)$ values received from its peer.

The $V(R)$ denotes the sequence number of the next in-sequence I frame expected to be received. The value of $V(R)$ is incremented by one with the receipt of each error-free, in-sequence I frame whose $N(S)$ equals $V(R)$.

The $N(S)$ is the send sequence number of transmitted I frames. Only I frames contain the $N(S)$. At the time that an in-sequence I frame is designated for transmission, the value of $N(S)$ is set equal to $V(S)$.

The $N(R)$ is the expected send sequence of the next received I frame. The value of the $N(R)$ contained with the Control field indicates that the data link layer entity transmitting the $N(R)$ has correctly received all I frames numbered up to and including $N(R) - 1$.

5.4.1.2 Poll/Final (P/F) Bit

All frames contain the Poll/Final (P/F) bit as illustrated in the Control field format of Figure 5.5. In command frames, the P/F bit is known as the P bit. In response frames, it is referred to as the F bit. When the P bit is set to 1, the sending data link layer entity is soliciting a response frame from its peer data link layer entity. The F bit set to 1 is used by a data link layer entity to indicate the response frame transmitted as a result of a poll command that is being responded to.

If a data link layer entity receives a SABME or DISC frame with the P bit set to 1, it sends out a response frame of UA or DM with the F bit set to 1. If it receives an I frame, RR, or RNR with the P bit set to 1, it sends out a response frame with the RR or RNR frame containing the F bit set to 1.

The P/F bit is used only for acknowledged multiple frame information transfer. For unacknowledged information transfer, the P/F bit is not used and is set to 0.

5.4.2 Frame Types

Frame types are either commands or responses. They can be used by either the user or the network data link layer entities.

The Control field identifies the type of frame. Three types of Control field formats are defined: numbered information (I format), supervisory functions (S format), and unnumbered information transfers and control functions (U format). These are explained in detail in the sections that follow.

5.4.2.1 Information Frame

```
Information (I)
```

The function of the I command frame is to enable information transfer across the established data link connection. The I command transfers sequentially numbered frames containing Information fields provided by layer 3. This command is used in the multiple frame operation on point-to-point data link connections.

5.4.2.2 Supervisory Frames

```
Receiver Ready (RR)
```

The RR supervisory frame is used by a data link layer entity to indicate that it is ready to receive an I frame. It is also used to acknowledge previously received I frames numbered up to and including $N(R) - 1$. The RR also clears out a busy condition that was indicated by an earlier transmission of a RNR (Receiver Not Ready) frame by that same data link layer entity.

The RR command with the P bit set may be used to ask for the current status of its peer data link layer entity.

```
Receiver Not Ready (RNR)
```

The RNR frame indicates a busy condition. The data link layer entity indicates to its peer entity that it is temporarily unable to accept additional incoming I frames. The value of $N(R)$ in the RNR frame acknowledges I frames numbered up to and including $N(R) - 1$.

The RNR command with the P bit set may be used to ask for the current status of its peer data link layer entity.

```
Selective Reject (SREJ)
```

The SREJ frame is used to request retransmission of the single I frame numbered $N(R)$. The $N(R)$ of the SREJ frame does not indicate acknowledgment of any I frames.

The SREJ frame usage is the preferred way of handling errored frames over a lossy medium like the airlink. This provides an efficient way to recover from occasional loss of frames in a sequence.

5.4.2.3 Unnumbered Frames

```
Set Asynchronous Balanced Mode Extended (SABME)
```

The SABME unnumbered command is used to place the addressed user or network side into modulo 128 multiple frame acknowledged operation. No Information field is sent with the SABME command. A data link layer confirms acceptance of an SABME command by transmission at the first opportunity of a UA response. The transmission of an SABME command indicates the clearance of all exception conditions.

```
Disconnect Mode (DM)
```

The DM unnumbered response is used by a data link layer entity to report to its peer that the link layer is in such a state that multiple frame operation cannot be performed. No Information field is sent with the DM response.

```
Disconnect (DISC)
```

The DISC unnumbered command is used to terminate the multiple frame operation. No Information field is permitted with the DISC command. The receiving data link layer entity confirms the acceptance of the DISC command by transmission of a UA response.

```
Unnumbered Information (UI)
```

UI commands are used to send information using the unacknowledged information transfer service of the data link layer. UI command frames do not carry a sequence number and therefore, the UI frame may be lost without notification to the layer 3.

```
Unnumbered Acknowledgment (UA)
```

The UA unnumbered response is used by a data link layer entity to acknowledge the receipt and acceptance of the mode-setting command like SABME or DISC. No Information field is permitted with the UA response. The transmission of the UA response indicates the clearance of any busy condition that was reported by a possible earlier transmission of an RNR frame by that same data link layer entity.

```
Frame Reject (FRMR)
```

The FRMR response may be received by a data link layer entity as a report of an error condition not recoverable by retransmission of the identical frame. An Information field that immediately follows the Control field and consists of five octets is returned with this response and provides the reason for the FRMR response.

```
Test (TEST)
```

This command frame is useful in conducting loopback testing between peer data link layer management entities. No sequence numbers are contained within the Control field of a TEST frame. An Information field is also included in the frame.

The contents of the Information field can be chosen by the link layer management entity initiating the loopback. The link layer entity responding to a loopback test returns the Information field from the initiator.

```
Zap (ZAP)
```

The ZAP command frame is used by the network side link management entity to disable a user side data link entity for a period of time. The Address field of the ZAP command contains a point-to-point TEI. This command is most useful in control of malfunctioning M-ESs. On receipt of a ZAP command, the M-ES is not to transmit any data link layer frame for any reason for the time defined in the ZAP Information field.

5.5 LINK ESTABLISHMENT PROCEDURE

Figure 5. 6 illustrates the flows that are part of this procedure.

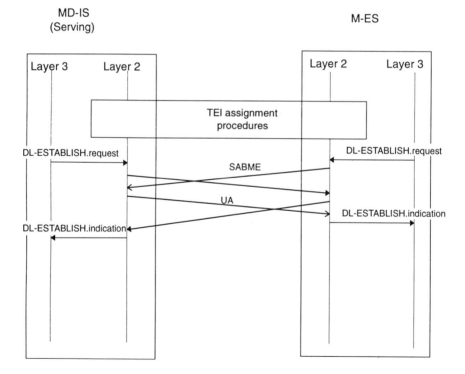

FIGURE 5.6 Link establishment procedure.

Layer 3 requests establishment of the multiple-frame operation with the DL_ESTABLISH.request primitive. The link layer performs the TEI administration procedures to enter the TEI-assigned state. Any outstanding I frames and any other DL_DATA.request primitives are discarded at this point.

The link layer then initiates a request for the multiple frame operation by transmitting the SABME command. All existing exception conditions are cleared, the retransmission counter is reset, and timer T200 is started.

The receiving data link layer entity, on receiving the SABME command shall do the following:

- Respond to the originating link layer entity with the UA response with the F bit set to 1.
- Set $V(S)$, $V(R)$, and $V(A)$ to 0.
- Enter multiple frame established state and inform the local layer 3.
- Clear any flow control and exception conditions.

At the originating link layer, on reception of the UA response with the F bit set to 1, the link layer entity will do the following:

- Reset timer T200.
- Set $V(S)$, $V(R)$, and $V(A)$ to 0.
- Enter multiple frame established state and inform the local layer 3.

5.6 SLEEP MODE

Since wireless portable devices typically operate from battery power, power conservation strategies are especially important. The link layer includes special procedures that aid in conserving battery power. The design of the link layer protocol supports a special *sleep* mode that the M-ES and the MD-IS link layer entities enter for a data link connection. In this mode of operation, the data link is intermittently active.

Figure 5.7 shows the exchanges between the M-ES and the MD-IS in support of the sleep mode.

The timer value T203 is negotiated as part of the TEI assignment set of exchanges between the M-ES and the MD-IS. After a period of inactivity at the link layer for a time equal to the predetermined T203, the sleep management procedures are executed at the M-ES and the MD-IS peer link layer entities. At the next expiration of the configured T204 interval, the MD-IS checks to see if any pending messages are lying at the MDLP layer destined for the sleeping M-ES. If there are any messages, the destination M-ES TEI is included in the list of TEIs that are part of the TEI notification message. The M-ES wakes up on each expiration of the T204 timer and waits for the TEI notification message to arrive. It receives and analyses this message to determine if its own TEI is part of the contents of this message. If it does

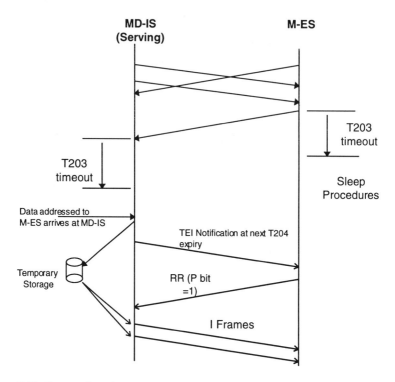

FIGURE 5.7 Sleep mode exchanges.

not find its own TEI there, it simply goes back to sleep. If it finds its own TEI to be a part of the received TEI notification message contents, it must send out a RR frame in order to inform the MD-IS that it is now ready to receive any stored frame lying at the link layer of the MD-IS. On receiving this RR frame, the MD-IS sends out all stored messages meant for this M-ES, according to the acknowledged data transfer model of operation of the MDLP.

Sleep procedures work best in a bursty data traffic environment with long periods of silence between rounds of data transfer for a pair of communicating data link layer entities.

5.7 BROADCAST DATA TRANSFER

Broadcast information transfer frames are directed to all user side endpoints on the channel stream. Only unacknowledged operation is possible in broadcast information transfer. Predefined, well-known TEI values are used for this operation.

The Broadcast TEI value of 0 is used for layer 2 management procedures such as identity request and assignment, TEI notification messages, identity check request and response messages, and identity remove messages.

The broadcast TEI value of 1 is used for layer 3 broadcast information transfer.

The link layer PDU that is used for broadcast data transfer is as follows:

< Broadcast TEI (0 or 1) > < UI format control field > < Layer 3 information field >

5.8 MULTICAST DATA TRANSFER

Multicast information transfer is a point-to-multipoint service available from the network side. Standard TEI management procedures are used to set up point-to-point data link connections for each of the multicast group M-ESs. There are special multicast NEI (network entity identifier—typically the layer 3 address, for example, the IP address) registration procedures (part of the layer 3 procedures) that are executed to allow multicast NEI authentication. Information transfer is one-way (MD-IS to the M-ES) and this uses the broadcast TEI value of 1 on each channel stream that has at least one M-ES that belongs to the multicast address group.

Further details on multicast data transfer are part of Chapter 8.

5.9 TYPICAL FLOW SCENARIOS

This section will show some typical operational scenarios and functional walk-throughs in the form of flow diagrams between the two peer MDLP entities at the M-ES and the MD-IS.

5.9.1 Multiple Frame Operation Establishment

The layer 2 management entity at the M-ES initiates a TEI request message (MDL_UNITDATA.Request) for the MDLP to send out on the acquired channel (the channel acquisition procedures are detailed in Chapter 4). When the layer 2 management entity receives a TEI assignment message (MDL_UNITDATA.indication) from the MD-IS (forwarded by the MDLP layer), it informs MDLP that the M-ES is now in TEI-assigned state. This triggers the MDLP to send out a SABME command to attempt to bring up the multiframe reliable data link protocol. Note that either side (M-ES side or the MD-IS side) can send out SABME and the other side must respond with a UA response frame. When MDLP receives the UA response to the SABME, it informs the SNDCP (using MDL_Establish) that the multiframe data link has been

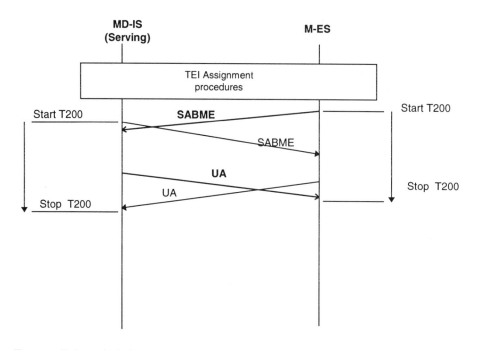

FIGURE 5.8 Multiple frame operation establishment flows.

established. Following this, the key exchange procedures (Chapter 6) and registration procedures (Chapter 8) are executed.

In Figure 5.8, two sets of SABME-UA flows are shown, one per SABME initiated by each of the two data link layer entities.

5.9.2 Acknowledged Data Transfer

Figure 5.9 illustrates the message flows used in this form of data transfer.

This figure assumes that an occasional I frame is lost over the airlink. Therefore, there are some sequences of SREJ or T200 retransmissions depicted in the message flows.

The use of a T205 acknowledgment delay timer is illustrated in this figure. This timer helps the data link layer entity to acknowledge multiple I frames, thus achieving some optimization for the usage of the air bandwidth.

5.9.3 Unacknowledged Data Transfer

Figure 5.10 illustrates the message flows used in this form of data transfer.

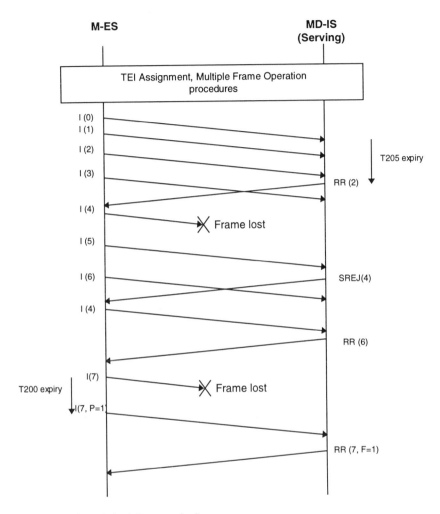

FIGURE 5.9 Acknowledged data transfer flows.

5.9.4 Intra-Area Cell Transfer

Figure 5.11 illustrates the message flows used in this form of data transfer.

The M-ES uses some of the RRM-generated parameters along with its own implementation-dependent algorithms and cell transfer criteria, and decides to exit the current cell. It then goes into execution of the procedures for selecting another cell within the same MD-IS coverage area. Once an appropriate cell is found, it enters that new cell. At this time, it sends out a RR with the P bit on (RR – P) frame from its

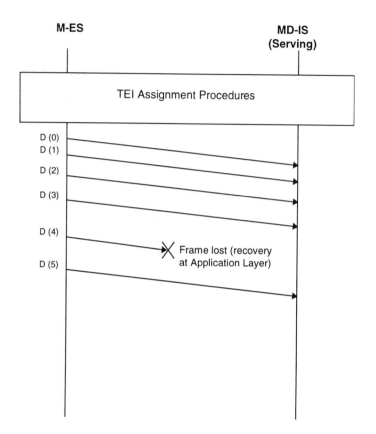

FIGURE 5.10 Unacknowledged data transfer flows.

MDLP layer. It waits for a RR with the F bit on (RR – F) frame from the MDLP entity at the current serving MD-IS. The receipt of the RR-F link layer PDU confirms that the MD-IS link layer and its associated management entity have kept track of the cell transfer operation of this M-ES. This in turn leads to the location information base update at the serving MD-IS (further detailed in the mobility management sections of Chapter 8).

If acknowledged data transfer was in progress at the time the cell exit and entry decision is taken by the M-ES, the receipt of any I frame at the serving MD-IS indicates the new cell location of the M-ES.

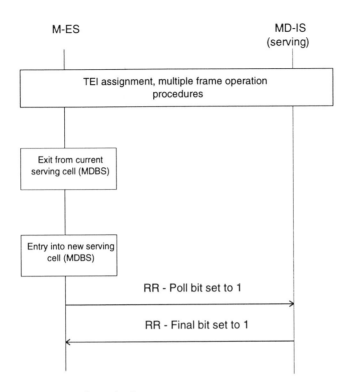

FIGURE 5.11 Intra-area cell transfer flows.

References

[1] RFC 791: Postel, J. B., Internet Protocol, September 1981.

[2] ISO-8473: Information Processing Systems—Data Communications Protocol for Providing the Connectionless-ModeNetwork Service and Provision of Underlying Service.

[3] ISDN User-Network Interface Data Link Layer Specification. International Telegraph and Telephone Consultative Committee, 1988. Recommendation Q.921.

[4] Data Communication—High-level Data Link Control Procedures—Consolidation of Elements of Procedure. International Organization for Standardization and International Electrotechnical Committee, 1984. International Standard 4335.

[5] CCITT-Q.920: ISDN User-Network Interface Data Link Layer—General Aspects. Nov. 1988. International Telegraph and Telephone Consultative Committee, 1988. Recommendation Q.920.

[6] ISO-7498: Information Processing Systems—Open Systems Interconnection: Basic Reference Model. International Organization for Standardization and International Electrotechnical Committee, 1984. International Standard 7498.

CHAPTER 6
▼▼▼

SUBNETWORK-DEPENDENT CONVERGENCE PROTOCOL LAYER

6.1 NEED FOR THE SUBNETWORK PROTOCOL LAYER

The network layer protocol data units (PDUs) originated by a variety of layer 3 protocols have to get transmitted to their peer entities across the CDPD airlink. This function requires the layer 3 PDUs to get transformed into a format appropriate for the link layer. The subnetwork-dependent convergence protocol layer (SNDCP) performs this function of packet transformation into segments that are handed over to the data link (the MDLP) layer.

The subnetwork-dependent convergence functions fill the gap between the different assumptions made by the multiple network layer protocol entities and the services actually provided by the single data link layer. These functions undertake the job of multiplexing data units generated by several network layer protocols into a single sequence of transmission. This single sequence of transmission is then handed over to the link layer. On the reception front, these functions provide the function of demultiplexing a sequence of data units received from the single data link layer. The received data units are assembled into network-level packets meant for delivery to the destination layer 3 protocol entities.

127

6.1.1 Multiple Network Protocols That Use the Link Layer

The CDPD network is a multiprotocol connectionless network. Sharing a single data link connection requires that different network layer entities and protocols be identifiable. The identification value is determined by the layer 3 entities and is analyzed by the SNDCP entity. The parameter that achieves this identification is termed the network layer protocol identifier (NLPI). Figure 6.1 illustrates network layer protocols and the NLPIs.

The layer 3 protocols that originate packets to be sent across the CDPD airlink include those listed in Table 6.1 (along with the respective NLPI parameter values).

TABLE 6.1
NLPI Values

NLPI	Network Layer Protocol Entity
0	Mobile network registration protocol
1	Security management entity defined as an application to provide data security across the airlink
2	Connectionless network protocol (CLNP) [1]
3	Internet protocol (IP) [2]

Of the above, the first two are meant for transport of user information, and the other two are applications that support the CDPD network and help make it a commercial service. Since CLNP is not widely supported in applications, IP is the key protocol used today for wireless messaging applications. With more applications around the corner, the list of protocols mentioned above is likely to increase. The SNDCP will then be serving a greater variety of network layer protocols.

6.2 OVERVIEW OF SNDCP

An SNDCP can be essentially modeled as supplying mapping functions between network-independent protocols and network-dependent services.

6.2.1 Model of Operation

As explained in the overall packet flow scenario in Chapter 4, the network layer packet goes through the following steps at the SNDCP layer. These steps are undertaken either at the M-ES or at the MD-IS, depending on the direction of information transfer:

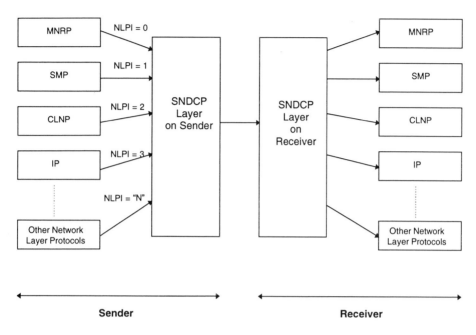

FIGURE 6.1 Network layer protocols and the NLPIs.

- The packet header is compressed (optional);
- The packet is compressed (optional);
- The packet is segmented, and each segment gets a segment header;
- Each segment is encrypted.

The SNDCP functions provide two service levels:

- Unacknowledged (connectionless mode) service class;
- Acknowledged (reliable connection mode) service class.

The protocol functions operate in a sequential manner, as shown in Figure 6.2 and Figure 6.3.

The processing undertaken at the SNDCP layer, for transmitting a network layer packet comprises the following:

1. Header compression function, which removes protocol control information that could be derived from saved PDU headers from an earlier packet with the same address pairs. This would target TCP/IP Header fields in an IP datagram and the CLNP header in an OSI CLNP datagram. This function is only executed in the information transfer over the acknowledged data link service.

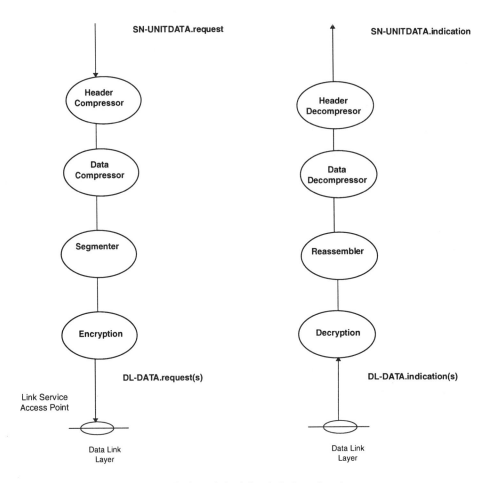

FIGURE 6.2 SNDCP as a user of acknowledged data link class of service.

2. Data compression function, which optionally encodes a string of user data octets in compressed format. This is done using the V.42 bis dictionary. This function is only executed in the information transfer over the acknowledged data link service.

3. Segmenting function, which generates one or more segments that are ready to go down to the link layer. This function is executed in both the unacknowledged and acknowledged data link layer classes of service.

4. Encryption function, which encrypts the output of the segmenting function. This function is only executed in the information transfer over the acknowledged data link service.

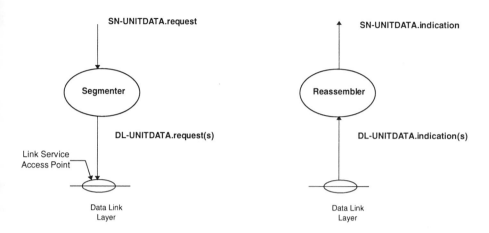

FIGURE 6.3 SNDCP as a user of unacknowledged data link class of service.

The processing undertaken at the SNDCP layer for receiving frames coming in from the data link layer comprises the following:

1. Decryption function, which decrypts the Data Segment field of a received SNPDU into plain digital data. This function is only executed in the information transfer over the acknowledged data link service.
2. Reassembly function, which assembles an ordered sequence of SNPDUs into a single PDU. This function is executed in both the unacknowledged and acknowledged classes of service.
3. Data decompressor function, which recovers strings of user data octets from received codewords using the V.42 bis functions. This function is optional and is executed in the acknowledged class of service.
4. Header decompressor function, which regenerates the protocol control information from earlier saved PDU headers. It then signals the network layer that the packet is ready for delivery to the local layer 3. This function needs to be used only for the acknowledged class of service.

6.2.2 SN-DATA Protocol Data Unit

The SN-DATA PDU is conveyed over the acknowledged data link service in the DL-Userdata field of a DL-DATA service primitive (this service primitive is explained in Chapter 5).

Figure 6.4 shows the format of an SN-DATA protocol data unit.

Bit	8	7	6	5	4	3	2	1
Octet 1	M	K	Comp. Type		NLPI			
Octet 2 - N	Network Layer Data (One Segment)							

MSBit   LSBit

FIGURE 6.4 Format of an SN-DATA protocol data unit.

In the figure, the letters show the following:

- M: more segments (1 = more segments, 0 = last segment);
- K: key sequence number (0,1);
- Comp: compression type—the values are specific to the type of network layer protocol in use (Section 6.3.1);
- NLPI: network layer protocol identifier (layer 3 protocols as defined in Section 6.1.1);
- Data: one segment of data meant for the data link layer (the MDLP). Maximum size is 128 octets.

6.3 HEADER COMPRESSION

With a view to preserve the precious bandwidth available in a wireless messaging environment, it is desirable to remove redundant protocol header information during transmission of a sequence of NPDUs transferred between the same pair of source and destination addresses. The compression protocol used is specific to the particular network and the transport layer protocols in use.

The general concept that drives protocol header compression is based on the typical pattern of usage. This is derived from a predictive pattern of usage and is based on indications that a relatively large number of packets are likely to be exchanged between a source and destination address pair, with only a very small amount of change in the protocol Header fields from NPDU to NPDU. The sender and receiver can maintain copies of the last NPDU header transmitted per source and destination pair and transmit only the differences along with a small address pair identifier. Since the compression function relies on detecting PDU loss and subsequent retransmission to resynchronize the compressor and decompressor after a transmission failure, compression cannot be reliably applied to NPDUs transmitted on a point-to-multipoint broadcast data link.

6.3.1 TCP/IP Protocol Header Compression

The protocol described by [3] is used for encoding and compressing the standard 40-octet TCP/IP protocol header to an average of 3 octets. A standard uncompressed TCP/IP header is reproduced in Figure 6.5. The fields that are shaded are the ones

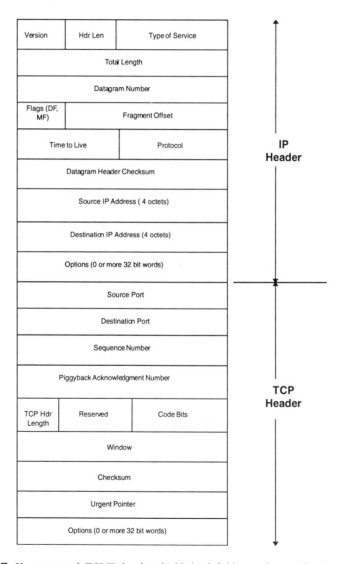

FIGURE 6.5 Uncompressed TCP/IP header—highlighted fields can be usually eliminated in subsequent packets.

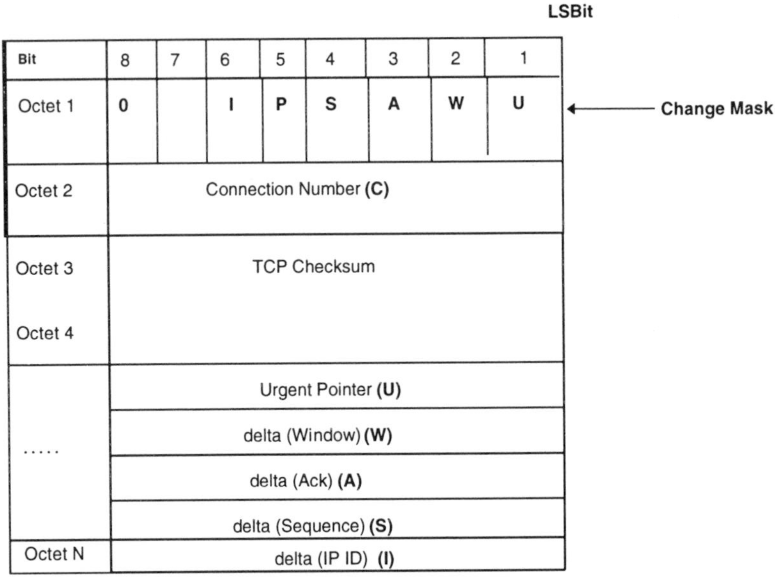

LSBit

Bit	8	7	6	5	4	3	2	1	
Octet 1	0		I	P	S	A	W	U	← Change Mask
Octet 2	Connection Number **(C)**								
Octet 3	TCP Checksum								
Octet 4									
.....	Urgent Pointer **(U)**								
	delta (Window) **(W)**								
	delta (Ack) **(A)**								
	delta (Sequence) **(S)**								
Octet N	delta (IP ID) **(I)**								

FIGURE 6.6   Compressed TCP/IP header.

that can be eliminated if the corresponding Protocol field does not change from the previous PDU. The replacement compressed TCP/IP protocol header, shown in Figure 6.6, indicates which of the fields have been eliminated from the ones shown in Figure 6.5. It can be noticed that the Connection Number field of the compressed protocol header is limited to one octet, thereby allowing 256 possible address pair associations simultaneously. Although this is possible, the CDPD specification limits the range of permitted connection numbers from 0..15 to reduce memory requirements on the SNDCP function.

It may seem at first glance that the savings associated with the compression function are not so significant. However, this is not the case, as a lot of exchanges between the communicating end systems are short and bursty in most applications. Take the case of a remote interactive telnet from an M-ES to a F-ES. In this session, the user at the M-ES will typically type a character or a line at a time. Depending on the TCP/IP protocol suite implementation, each character may get transmitted across the airlink as a separate packet. This would mean that 40 octets of TCP/IP header information goes along with a single character of application-specific (telnet in this case) data. The usage efficiency of the air bandwidth can be significantly enhanced by the TCP/IP header function that will bring this down to 3 to 5 octets of header information per packet of application-specific data for the second and subsequent PDUs, after the initial exchange of seed or resynchronizer data packets that are uncompressed. This would translate into improved response time for the user of the telnet application.

In the case of a file transfer (FTP) application that also uses TCP/IP, the compression function will significantly improve the time taken to complete the transfer between the two end systems. This is due to the fact that the file transfer is a number of packets going in one direction with the same TCP endpoints. As the number of octets that can fit into one link layer packet is limited, if some of those octets get eliminated due to the compression function being used, the number of octets (and therefore, total number of packets) that have to be sent over the air are considerably reduced.

6.3.2 CLNP Protocol Header Compression

The protocol of [3] has been adapted for the specifics of encoding of the [1] protocol. In this case, no assumptions are placed on the transport layer. The compression operates only on the layer 3 CLNP header.

Figure 6.7 shows a typical CLNP header for a data PDU. This is a minimum of 57 octets. During an exchange of information between an address pair of {source, destination}, many of these fields are likely to stay constant or change only by small amounts from PDU to PDU.

The fields that stay constant during the lifetime of an association between two layer 3 address pairs and therefore never need to be transmitted in a compressed header are as follows:

- Network Layer Protocol Identifier field;
- Version field;
- Destination Address Length Indicator field;
- Destination Address field;
- Source Address Length Indicator field;
- Source Address field.

In addition to the above fields, the Segment Length field is not required as the data link layer will indicate the length of a received frame. The Lifetime field is redundant if the receiver is the M-ES or has the initial first-hop value if the receiver is the MD-IS. The Header Checksum field is redundant because the data link layer entity provides its own error detection mechanism.

These steps eliminate 49 octets of protocol control information from the compressed CLNP header.

Figure 6.8 shows the format of a compressed CLNP NPDU.

In the best case, communication between a single pair of NSAP addresses, the compression function will replace the 57-octet CLNP header with a 1-octet compressed header for the second and subsequent NPDUs exchanged between the communicating end systems.

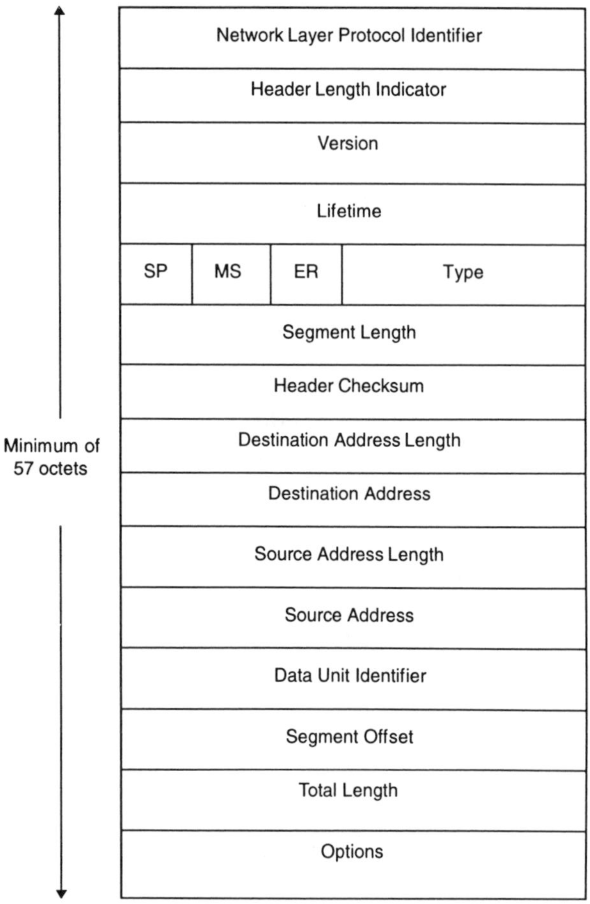

Network Layer Protocol Identifier			
Header Length Indicator			
Version			
Lifetime			
SP	MS	ER	Type
Segment Length			
Header Checksum			
Destination Address Length			
Destination Address			
Source Address Length			
Source Address			
Data Unit Identifier			
Segment Offset			
Total Length			
Options			

Minimum of 57 octets

FIGURE 6.7 Uncompressed CLNP header.

6.4 DATA COMPRESSION

The data compression protocol used is [4]. This compression function increases throughput over the airlink by encoding data prior to transmission such that strings of user data octets are represented by a sequence of codewords in fewer bits.

The SNDCP layer provides this data compression function to all network layer protocols that are supported. The data compression function relies on an underlying reliable data link and can therefore be used only on an acknowledged data link class of service.

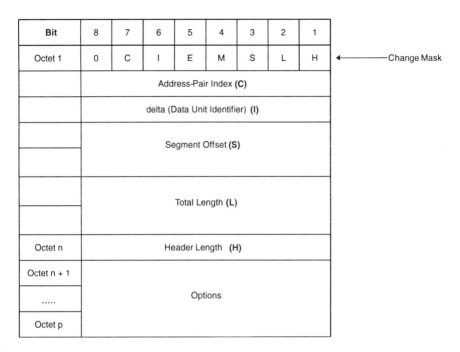

FIGURE 6.8 Format of a compressed CLNP NPDU.

6.4.1 Model of Operation

The use of the data compression function and its associated parameters is negotiated by the management entity at initial data link creation, via the TEI assignment procedures. These values remain in effect for the duration of the data link connection. The parameters that are negotiated are as follows:

- P_0—specifies whether compression is to be used and in which direction (forward channel only, reverse channel only, or both directions).
- P_1—specifies a proposed value for the total number of codewords. This is typically a power of 2 for best performance of the algorithm. The range for this parameter is between 512 and 8,192 codewords.
- P_2—specifies a proposed value for the maximum string length, within the range of 6 through 250.

The compression and decompression functions work according to the model shown in Figure 6.9.

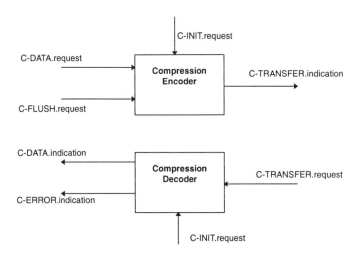

FIGURE 6.9   Model of V.42 bis compression functions.

The data compression function gets initialized after successful negotiation of the data compression parameters and on establishment of multiple frame operation.

In the transmit direction, each subnetwork service data unit (SNSDU) gets encoded by the data compressor according to the data compression dictionary implemented as a tree structure, as specified in V.42 bis. This encoding is done subsequent to protocol header compression and prior to segmentation.

In the receive direction, upon reassembly of a complete SNPDU sequence, the data decompressor decodes the data and delivers it to the protocol header decompressor function.

6.5   SEGMENTATION AND REASSEMBLY

The frames received at the SNDCP layer entity can be larger than what the data link layer can take as the maximum size data unit. The segmentation function at the transmitting entity and the reassembly function at the receiving entity achieve the mapping of these large packets into an ordered sequence of one or more SN-DATA or SN-UNITDATA PDUs, depending of whether the underlying data link service class is acknowledged or unacknowledged.

6.5.1   Segmenting and Reassembly Over the Acknowledged Class of Service

Segments generated in the sequence, except the last one, have the More Segments (M) field set to 1. The last (or if this was the only segment) SN-DATA PDU in the sequence has the M field set to 0. The receiving SNDCP entity reassembles a sequence of SN-DATA PDUs into a single SNSDU. An SNSDU does not get processed any further by SNDCP (functions like decompression) unless all the segments of the SNSDU are received.

If the underlying data link connection at the MDLP layer fails, the transmitter discards any unsent segments of a partially transmitted sequence. The receiver discards any partially assembled sequence.

Failure of the underlying data link connection is determined by any of the following events:

- DL-RELEASE.indication;
- DL-ESTABLISH.indication;
- DL-ESTABLISH.confirm.

An example of this kind of segmentation and reassembly would be that of the end system hello (ESH) packet sent out by the mobile network registration protocol (MNRP) layer 3 entity. This packet with all of its options included is larger than the maximum size of the size acceptable to the data link layer.

An ICMP ping message, of data length greater than 100 bytes, from the M-ES to an F-ES is another example of an SNSDU requiring segmentation and reassembly.

6.5.2 Segmenting and Reassembly Over the Unacknowledged Class of Service

All segments generated from a single SNSDU (these segments get mapped onto SN-UNITDATA PDUs), are assigned a sequence identifier, modulo 16. This identifier is incremented each time a complete SN-UNITDATA sequence is transmitted. Each PDU segment in the sequence is sequentially numbered, starting at 0. Four bits are available to assign the segment number within the sequence. This means that a complete sequence of SN-UNITDATA PDUs can consist of 1 to 16 consecutive segments. Segments generated in the sequence, except the last one have the More Segments (M) field set to 1. The last (or, if this was the only segment) SN-UNITDATA PDU in the sequence has the M field set to 0.

Failure to receive the complete SNSDU, consisting of its various segments, is determined by the following events:

- Expiration of the reassembly timer. This timer protects the receiver from the loss of a series of segments and possible sequence identifier wraparound.
- Receipt of an SN-UNITDATA PDU with a different sequence identifier.
- Intra-area or interarea cell transfer.

Any partially assembled sequence is discarded on a receive failure event.

6.6 ENCRYPTION AND DECRYPTION

Encryption and decryption are applied by SNDCP only to point-to-point data link connections using the acknowledged data link service. The broadcast data link connections or point-to-point data link connections using the unacknowledged mode service do not use the services of encryption provided by the SNDCP entity. The user of the SNDCP service can however, apply end-to-end encryption.

Encryption requires the exchange of a key sequence number in encrypted packets. This is a 1-bit sequence number (K) parameter illustrated in the SN-DATA PDU in Figure 6.4, shown earlier in this chapter.

Security being a very complex set of issues in computing and wireless communications, security services must be capable of supporting a wide range of applications that extend beyond the confines of SNDCP or MDLP. The security management entity (SME) defines a set of services offered between the M-ES and the MD-IS that makes use of the services provided by SNDCP. The peer-to-peer procedures for exchange of encryption keys and encrypted information are the subject matter of the SME.

References

[1] ISO-8473: Information Processing Systems—Data Communications Protocol for Providing the Connectionless-Mode Network Service and Provision of Underlying Service.
[2] RFC-791: Postel, J. B., Internet Protocol, September 1981.
[3] RFC-1144: Jacobson, V., Compressing TCP/IP Headers for Low-Speed Serial Links, February 1990.
[4] CCITT-V.42 bis specifications.

CHAPTER 7
▼▼▼

RADIO RESOURCE MANAGEMENT

As an overlay network on AMPS, the radio end of the CDPD architecture is closely tied to the behavior and the operational characteristics of the RF subsystem of the analog mobile phone system (AMPS) system. This chapter outlines the key architectural elements of the CDPD architecture that manage and control the radio resources. This will serve as a foundation for describing the radio resource management strategies employed to use the available RF channels effectively to provide a data channel for moving end systems.

CDPD technology is dependent on the existence of the infrastructure equipment, RF channel availability, and allocation strategies of the AMPS. Hence the popularity and survivability of CDPD is closely associated with AMPS technology and its market share with other competing technologies. Especially when the CDPD system is configured to share frequencies with the AMPS system, the efficiency and the effectiveness with which the radio resources are managed will be key factors in determining the success of CDPD as a viable technology for reliable, high-throughput wireless data transfer.

7.1 CELL-BASED NETWORK ARCHITECTURE

Cellular architecture is fundamental to the AMPS system and second/third-generation wireless digital networks that have potential to support CDPD technology.

7.1.1 Omni/Sectored Cells

In a mobile telephone system, the cellular concept provides a convenient framework for the design of the infrastructure of the network and frequency assignment from a limited available RF channel set. In a system such as AMPS, where a limited RF channel set is to be used to obtain traffic capacity in a large geographic area that far exceeds the number of available RF channels, the following considerations simplify designing cellular networks with omni antennas:

- The reuse requirement of a specific RF channel requires that its use is confined to an identifiable geographic area.
- Assuming a flat terrain, a simple omnidirectional antenna placed on a high tower will provide a coverage area with an approximately circular pattern. The coverage area represents the area bounded by a contour of constant signal level that can provide acceptable service quality. In practice, uneven terrain and obstacles such as tall buildings and hills produce an irregular boundary of radio coverage.
- Designers choose a hexagonal shape as a suitable approximation to a cell as it closely fits to a circle, and it can be used repetitively to cover larger geographic areas without any overlap.

However, as shown in Figure 7.1, in most current cellular systems, cell sites are sectored with three faces, each face having a transmit and receive antenna, which are directional, covering 120 degrees. The antenna's main front lobe will form the two edges of the associated hexagonal cell. Directional sites provide less co-channel interference and thereby closer spacing between two co-channel sites can be tolerated. This provides greater capacity for a given geographic area if a sectored cell configuration is chosen for radio coverage.

7.1.2 Adjacent Cells

To allow implementation of efficient handoff mechanisms, network configurations require defining neighbor cells for each cell in the network. Cell selection algorithms in the M-ESs for handing off when an M-ES continuously moves to different physical locations depend on the number of different adjacent cells that have been defined for the current cell. Figure 7.2 illustrates adjacent (neighbor) cell relationships.

In sectored cells, sectors of a cell are defined more specifically as face neighbors. As will be seen later, this relationship is used for more efficient cell transfer strategies when an M-ES moves across sector boundaries.

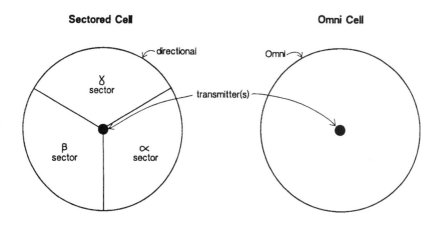

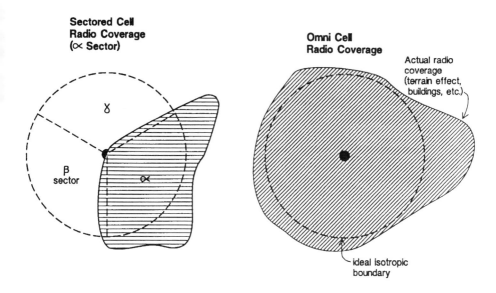

FIGURE 7.1   Omni and sectored cell configurations.

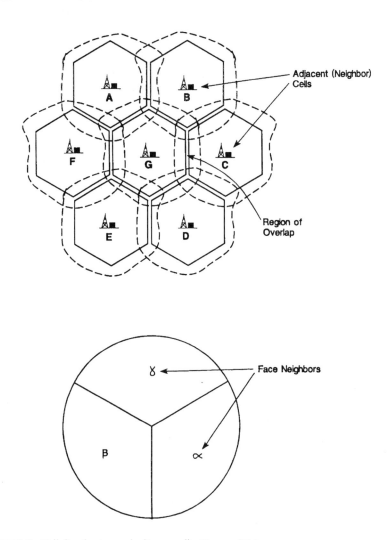

FIGURE 7.2   Cell distribution and adjacent cells. (*Source:* [1].)

7.1.3 Frequency Reuse and Cell Clusters

Since a cell may require multiple RF channels and adjacent cells require other frequencies, strategies have to be developed to group cells and allocate the given block of frequencies to the group, each getting an equal share of the frequencies in the ideal case. The geometry of the group of cells must be such that the groups can be used as a *macro* building block to cover any network topology. Omni 3, 4, and 7 site clusters (groups) and sectored 3/9, 4/12, and 7/21 cell clusters are illustrated in Figure 7.3.

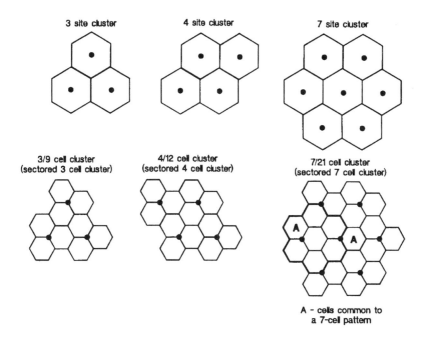

FIGURE 7.3 Cell clusters. (*Source:* [1].)

A set of several frequencies are used in one cell and other mutually exclusive sets are used in other cells in the cluster. This arrangement forces one to examine the criteria as to when a frequency can be reused again. Figure 7.4 provides an example of the popular 7-cell cluster arrangement, with six clusters surrounding a cluster in the middle. In the middle of each cluster is a base station that must have the same frequency set if the network is being built with the 7-cell cluster as the basis.

It can be shown that if the cluster size is N and the cell radius is R, the reuse distance D is given by

$$D = R \, (3N)^{1/2}$$

The cluster sizes need not be the same throughout the network. Cells can be split to provide increased capacity (number of subscribers per square area) towards city centers. Figure 7.5 shows such a plan with representative cluster sizes of 4, 7, and 12.

7.1.4 Adjacent Areas

Each MD-IS and the MDBSs that it supports in a CDPD network belong to an *area*, as illustrated in Figure 7.6. When an M-ES moves from a cell to another cell that is

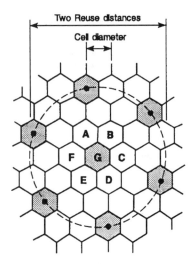

FIGURE 7.4 Frequency reuse distance. (*Source:* [1].)

supported by a different MD-IS, the M-ES is expected to reregister with the new MD-IS.

The reregistration requirement during interarea cell transfers is an overhead the M-ESs will attempt to avoid when selecting RF channels. If an acceptable cell is available within the same area, the intra-area transfer is given priority.

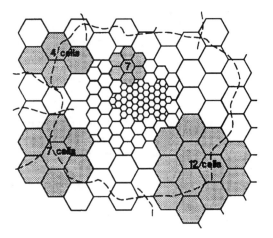

FIGURE 7.5 Typical cell layout plan for a city. (*Source:* [1].)

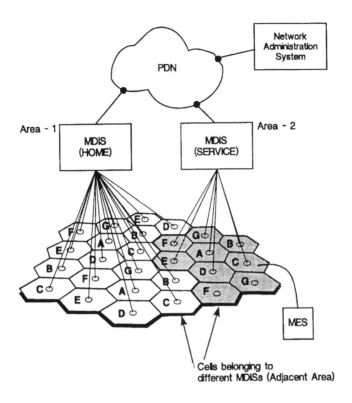

FIGURE 7.6 MD-ISs and adjacent areas.

Each M-ES has a home MDIS, the significance of which will be detailed when the mobility aspects of M-ESs are covered in the next chapter.

7.1.5 CDPD Frequency Pools and AMPS System

CDPD deployment is normally carried out in regions where the AMPS system is already in operation. Operational AMPS systems will have an established frequency plan. However, the frequency plans continually change if the network is being expanded by adding cell sites or splitting existing cells. A complete network-wide reassignment of frequencies is not uncommon. In this scenario, carriers have several alternatives when assigning frequencies for an overlaid CDPD system.

- *Dedicated frequencies.* If one or more spare frequencies are available that are not required for use in the AMPS system, these frequencies can be assigned for use by CDPD. These are called *dedicated* frequencies. The CDPD network can assume that these frequencies will not be used by any non-CDPD system.

- *Shared frequencies.* These frequencies are shared by both the voice system and the CDPD system. The CDPD system is expected to employ mechanisms to detect the use of these frequencies by non-CDPD systems. While CDPD is using an RF channel, if non-CDPD use is detected, the CDPD system is expected to relinquish using the RF channel within 40 ms.

In multiple channel streams per sector configurations, a mix of dedicated and shared frequencies can be used within a cell. The RF channel selection algorithms are expected to give priority to dedicated channels over shared channels.

When shared frequencies are used in a CDPD system, one of two methods can be employed to manage their use.

- *Sniffer hardware.* Hardware can be added to the CDPD MDBS to continuously scan the shared frequencies to detect presence of energy. This status information can assist the channel allocation portion of the software. More importantly, the hardware also has to provide information on the shared frequencies that are currently used by the CDPD system; this information has to be provided at a faster rate so that the MDBS can perform a forced hop when voice activity is detected within the 40 ms required by the specifications.
- *Channel status protocol (CSP).* Close coordination between the voice and the CDPD system can be organized by implementing a channel status protocol between them. Messages exchanged between the two systems can avoid both systems choosing the same frequency together. This protocol has not been widely implemented in CDPD and AMPS systems yet.

7.1.6 Color Codes

Color codes in the CDPD system perform the same function as the supervisory audio tones (SATs) in the analog cellular system. In addition to facilitating the detection of co-channel interference, the CDPD color code also is used when an M-ES performs an interarea cell transfer.

7.1.6.1 *Group Color Code*

The CDPD specification defines the group color code as follows. Each cell in a group of cells may be assigned the same cell group color code if both of the following conditions apply:

- The cell is adjacent to another cell in the same group.

- No radiofrequency (RF) channel available for CDPD use in the cell is also available for CDPD use in any other cell in the group.

The concept behind the group color code is the same as the supervisory audio tones (SATs) used in the voice network. It helps the receiving element (cell or an M-ES) to distinguish between two transmitting elements (M-ESs or cells, respectively) that are using the same frequency.

The CDPD group is equivalent to a cell cluster in the voice network and is therefore associated with the frequency assignment strategy employed by a carrier. Frequency reuse patterns determine the cell groups. Frequency reuse refers to the use of radio channels on the same frequency to cover different areas that are separated from one another by sufficient distances so that co-channel interference is not significant enough to disturb the operation of the system. A set of cells that have mutually exclusive frequency sets and are geographically contiguously located to satisfy the neighbor criterion above will be labeled as a group and will be given the same group color code. Since the group color code can have 31 distinct values (5 bits), it would not be a difficult task to assign the color codes so that no two cells transmitting on the same frequency within the hearing range of a single M-ES have the same color code.

The group color code is encoded in each RS block transmitted on the forward channel stream and is used by the M-ES to detect co-channel interference from a remote MDBS. Similarly, the group color code encoded in the first block of a burst transmitted on the reverse channel stream is used by the MDBS to detect co-channel interference from a remote M-ES.

7.1.6.2 Area Color Code

Area color code is a unique identifier of 3 bits in length assigned to an MD-IS. The area color code has to be unique only within MD-ISs that geographically cover areas adjacent to each other. This is assigned by the network service provider when defining the network topology and the geographic placement of the MD-ISs. Along boundaries of the networks operated by different service providers, coordination in assignment of the color codes may be necessary to ascertain smooth operation when M-ESs roam from one service provider's network to another's network.

The area color code is broadcasted in the cell configuration messages. In addition, the area color code (3 bits) and the cell group color code (5 bits) are also transmitted together as the first data byte of each Reed-Solomon (RS) block in the forward channel. The area color code allows an M-ES to identify during a cell transfer whether it is about to move into an area controlled by a different MD-IS. Since an interarea cell hop requires the M-ES to reregister, the area color code can be used by the M-ES to favor a cell that has the same area color code as the current cell the M-ES is currently tuned to in order to prevent the reregistration overhead.

7.2 RADIO RESOURCE MANAGEMENT: MAIN GOALS

The radio resource management entity (RRME) is the part of the layer management entity whose function is to manage the available radio resources for efficient and correct operation of the CDPD system. RRME entities with differing management responsibilities reside in both MDBS and M-ES to achieve the following major goals:

- As an overlay technology on AMPS, the CDPD cell boundaries should be made to coincide with that of the AMPS system.
- With the use of passive techniques, the CDPD transmissions should avoid interfering with transmissions of the voice network.
- In situations where dedicated channels cannot be allocated to CDPD, the CDPD system should be able to use the bandwidth available during the idle time of frequencies (not used by voice) for CDPD transmissions. Switching the frequency used by a channel stream (RF port) is called a *channel hop*.
- Scarcity of bandwidth necessitates the requirement for using in-band control, unlike the voice system where dedicated control channels are available.
- An M-ES must be able to acquire a channel for its use within an acceptable delay without any prior knowledge of the frequencies in use, and to continue to maintain connectivity and exchange information with its remote end in the presence of channel hops.
- A moving M-ES must maintain connectivity and continue exchanging information with its remote end as it moves across cell boundaries, across areas operated by the same or different network service providers. In this scenario, the M-ES must be able to account for terrain effects in its selection of the best radio channel to use.

CDPD system employs three components of radio resource management methodology that combine to achieve the above main goals:

1. The RRME that resides in the M-ES;
2. The RRME that resides in the MDBS;
3. The radio resource management protocol (RRMP). This encompasses a set of messages exchanged between the RRMEs in the MDBS and in the M-ES, and the associated set of procedures that facilitate coordinated execution of radio resource management.

The RRME is distributed in different components of the CDPD architecture, as illustrated in Figure 7.7.

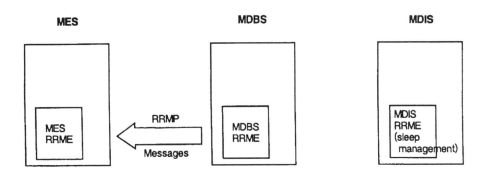

FIGURE 7.7 Different components of radio resource management.

7.3 RADIO RESOURCE MANAGEMENT PROTOCOL

The radio resource management protocol consists of a set of MDLP-level messages transmitted from the MDBS RRME to the M-ES RRME. The simplicity of this protocol is evident from the following:

- The RRMP messages are transmitted in the forward direction only. There are no RRMP messages from the M-ES RRME to the MDBS.
- The RRMP frames are carried as unnumbered information (UI) frames as part of the MDLP link, so that the RRME in the MDBS can easily insert the MDLP frames within the forward direction MDLP packets handled by the MDLP relay entity that will be normally resident in the MDBS.

The RRMP frames mainly carry network, cell, and channel identification information and the parameters necessary for the M-ES to perform its radio management functions. These frames are broadcast to all M-ESs. The following messages fall into this category:

1. Channel stream identification message;
2. Cell configuration message;
3. Channel quality parameters message;
4. Channel access parameters message.

A significant amount of configuration information is necessary to allow the MDBS RRME to construct these messages.

The switch channel message is the only other RRMP message sent as a command to the selected set of M-ESs.

The message formats are defined and the component fields in each messages are described. These parameters will provide the framework for describing the RRMEs and their respective functionalities.

7.3.1 Message General Format

The general format of an RRMP message, including the MDLP information is shown in Figure 7.8.

All RRMP messages have the TEI defined as 0, indicating that it is a broadcast message targeted to layer management entities. The messages are typed as commands and the effective address extension is 1 (no additional address bytes), making the first byte in the MDLP frame 0×03.

The frame type is unnumbered information, and in the extended mode of operation takes one byte, 0×03.

The Information field in the UI frame has two header bytes to define the following:

Byte-1
    Layer management entity identifier
    0×2A: RRME

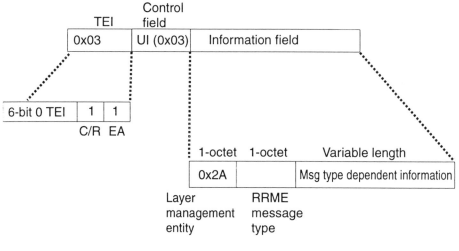

FIGURE 7.8 RRMP message format.

Byte-2
  RRME Message Type
    0×00: Channel stream identification
    0×01: Cell configuration
    0×02: Channel quality parameters
    0×05: Channel access parameters
    0×04: Switch channel message

At the MDLP frame level, the first five bytes of the different RRMP messages are, therefore:

Channel identification message:	03 03 2A 00...
Cell configuration message:	03 03 2A 01...
Channel quality parameters Msg:	03 03 2A 02...
Channel access parameters Msg:	03 03 2A 05...
Channel switch message:	03 03 2A 04...

7.3.2 Information in RRME Messages

7.3.2.1 Channel Stream Identification

The channel stream identification message is broadcast by the MDBS RRME in each channel stream at a periodic interval of, typically, 2–5 seconds. This message contains information for the M-ESs to identify the network, the cell, and the channel stream. The parameters in the message are shown in Figure 7.9 and described next:

- *Protocol version.* RRM protocol version. CDPD Specification 1.1 definition sets this value to 1.
- *RF channel type.* Defines whether the currently used RF channel on this channel stream is a CDPD dedicated frequency (bit set to 1) or a shared frequency that can undergo channel hops (0).
- *Channel capacity.* MDBS indicating to the listening M-ESs whether the current channel stream capacity is full (1) or not (0). M-ESs are not supposed to attempt to register on this channel if the bit is set.
- *Channel stream identifier.* A unique identifier (0–63) of a channel stream within a given cell.
- *Cell identifier.* This value is a globally unique identifier of a cell and consists of two 16-bit subfields made up of the service provider network identifier (SPNI) and the cell identifier. The cells within a single service provider network are assigned the same SPNI.

Bit	8	7	6	5	4	3	2	1
Octet 1	LMEI ($\approx 42_{10}$)							
Octet 2	Type = Channel Stream Identification (0)							
Octet 3	Protocol Version							
Octet 4	Dedicated	Capacity	Channel Stream Identifier					
Octet 5	Cell Identifier (= Service Provider Network Identifier + Cell number)							
Octet 6								
Octet 7								
Octet 8								
Octet 9	Service Provider Identifier							
Octet 10								
Octet 11	Wide Area Service Identifier							
Octet 12								
Octet 13	Power Product							
Octet 14	Max Power Level							

FIGURE 7.9 Parameters in the channel id message. (*Source:* [2]. © CDPD Forum.)

- *Service provider identifier.* The service provider identifier (SPI) uniquely identifies a licensed facilities-based cellular service provider offering CDPD services.
- *Wide area service identifier.* Service providers can define cooperating agreements to provide service over large geographic areas that may comprise several SPNIs. In such cases, the wide service area is identified by a unique wide service area identifier (WASI).
- *Power product.* Represents unsigned 8-bit value of the power product, expressed in decibels. The absolute value of the power product will be adjusted by adding –143 dB to keep the result always positive for the RRMP frame.
- *Maximum power level.* This is an unsigned value between 0 and 10 representing the maximum power level an M-ES can use for reverse transmission. The 0–10 levels are mapped to absolute power values.

7.3.2.2 Cell Configuration

The cell configuration message is broadcast on every channel stream by the RRME in the MDBS at periodic intervals of 5–10 sec. One cell configuration message pertaining to the current cell is broadcast on each channel stream. In addition, a cell configuration message for each of the neighbor cells configured for the current cell is also broadcast on each channel stream. These messages provide the M-ES information of

Bit	8	7	6	5	4	3	2	1
Octet 1	LMEI (=42₁₉)							
Octet 2	Type = Cell Configuration (1)							
Octet 3	Cell ID (= Service Provider Network Identifier + Cell number)							
Octet 4								
Octet 5								
Octet 6								
Octet 7	Face	0	Active Channel Streams			Area Color Code		
Octet 8	Reference Channel							
Octet 9								
Octet 10	ERP Delta							
Octet 11	RSSI Bias							
Octet 12	Power Product							
Octet 13	Max Power Level							
Octet 14	Dedicated	Reserved						
Octet 15	RF Channel Number 1							
•••••••	(RF Channel Numbers 2 to N-1)							
Octet 2N+12	Dedicated	Reserved						
Octet 2N+13	RF Channel Number N							

FIGURE 7.10 Parameters in the cell configuration message. (*Source:* [2]. © CDPD Forum.)

the RF channel configuration of the current and neighbor cells and other parameters that assist the M-ES during channel hops and cell transfers.

The parameters carried in the cell configuration message are shown in Figure 7.10 and described next.

- *Cell identification.* Unique cell id that includes the SPNI.
- *Face of a sectored cell.* This bit is 1, if the cell is a face of a sectored cell.
- *Active channel streams.* This number defines the number of active channel streams in a cell. Zero means CDPD is not active in that cell, and 7 means that there are 7 or more channel streams in that cell.
- *Area color code.* Denotes the area color code of that cell to facilitate the M-ES to identify cells supported by different MD-ISs.
- *Reference channel.* A valid RF channel number referring to a continued keyed channel in a defined cell that can be used in the best cell selection by the M-ES.

- *ERP delta.* This value defines the difference in the nominal effective radiated power (ERP) between the reference channel and the CDPD channels in that cell. The expected signal strength of a CDPD channel in a cell is calculated by subtracting the ERP delta value from the received signal strength indication (RSSI) value of the reference channel.
- *RSSI bias.* Effective RSSI value of an adjacent cell is calculated by adding the RSSI bias value to its expected RSSI value. A positive value for a neighbor cell shrinks the boundary of the current cell.
- *Power product:* The same quantity is broadcast as part of the channel stream id message.
- *Maximum power level.* The same quantity is broadcast as part of the channel stream id message.
- *RF channel type.* Dedicated or shared channel type.
- *RF channel number.* RF channel number (1–1,023). There will be as many pairs of these entries <channel type, channel number> as there are number of entries in the frequency pool of this cell.

7.3.2.3 Channel Quality Parameters

The channel quality parameters message is broadcast at periodic intervals, typically on the order of 60 seconds. This message contains information that assists the M-ES in determining the quality of the channel for continuing to use it for data transfer.

The parameters in the channel quality parameters message are shown in Figure 7.11 and described next.

- *RSSI hysteresis.* This parameter takes a value between –127 dB and 127 dB and is used as an adjustment to the current channel's RSSI value when select-

Bit	8	7	6	5	4	3	2	1
Octet 1	\multicolumn{8}{c}{LMEI (=42$_{10}$)}							
Octet 2	\multicolumn{8}{c}{Type = Channel Quality Parameters (2$_{10}$)}							
Octet 3	\multicolumn{8}{c}{RSSI HYSTERESIS}							
Octet 4	\multicolumn{8}{c}{RSSI SCAN TIME}							
Octet 5	\multicolumn{8}{c}{RSSI SCAN DELTA}							
Octet 6	\multicolumn{8}{c}{RSSI AVERAGE TIME}							
Octet 7	\multicolumn{8}{c}{BLER THRESHOLD}							
Octet 8	\multicolumn{8}{c}{BLER AVERAGE TIME}							

FIGURE 7.11 Channel quality parameters message format. (*Source:* [2]. © CDPD Forum.)

ing the best server channel. A positive hysteresis value forces the M-ES to favor the current channel, avoiding unnecessary flip-flops between two channels that have roughly equivalent signal strengths at the location of the moving M-ES.

- *RSSI scan time.* This represents the maximum time that may elapse before the M-ES RRME initiates an adjacent cell scan procedure.
- *RSSI scan delta.* This value defines the upper and lower ranges for the variation of the RSSI value of the current channel.
- *RSSI average time.* Averaging time for the RSSI value.
- *BLER threshold.* Channel exception condition is generated if the current averaged block error rate (BLER) exceeds this value, invoking an adjacent cell scan. BLER is defined as the ratio of the number of uncorrectable blocks compared to the total number of blocks received.
- *BLER average time.* Time for calculating moving average of the BLER value.

7.3.2.4 Channel Access Parameters

The channel access parameters message contain parameters related to the operation of the MAC layer. These parameters define the operational characteristics of multiple M-ESs sharing a given channel stream and the error-recovery procedures carried out by the MAC layer.

The parameters in the channel access parameter message are shown in Figure 7.12 and described next.

- *Maximum transmit attempts.* This parameter defines the number of retries the MAC layer should attempt to retransmit, either if channel busy when it tries to check the channel condition when it has data to transmit or an actual trans-

Bit	8	7	6	5	4	3	2	1
Octet 1	LMEI $(=42_{10})$							
Octet 2	Type = Channel Access Parameters (5_{10})							
Octet 3	MAX_TX_ATTEMPTS							
Octet 4	MIN_IDLE_TIME							
Octet 5	MAX_BLOCKS							
Octet 6	MAX_ENTRANCE_DELAY							
Octet 7	MIN_COUNT							
Octet 8	MAX_COUNT							

FIGURE 7.12 Channel access parameters message format. (*Source:* [2]. © CDPD Forum.)

mission is notified as decode failure in the forward channel. When the maximum transmission attempts fail, the MAC layer is expected to generate a channel exception condition triggering the mechanisms that perform frequency rescan.

- *Minimum idle time.* After a successful transmission of a burst, the M-ES is required not to access the reverse channel for a minimum period defined by the minimum idle time (in microslots units).

- *Maximum blocks.* Maximum number of blocks that can be transmitted in a burst by an M-ES.

- *Maximum entrance delay.* If an M-ES has data to transmit and senses the channel is busy, it defers, checking the channel again after a random number of microslots defined within 0 and the max entrance delay parameter.

- *Minimum count.* If a there is a collision in the reverse channel transmission detected by the decode failure flag, the M-ES backs off for a number of microslots selected within a uniformly distributed random number between 0 and $2^{\text{minimum count}} - 1$.

- *Maximum count.* During each successive failure to transmit, the random backoff delay parameter will be selected between 0 and $2^x - 1$, where x is the exponent used in the previous backoff calculation incremented by 1. Maximum count defines the upper limit to the incrementing exponent x. Thus, the valid range of x is

$$\text{minimum count} < = x < = \text{maximum count}$$

7.3.2.5 Switch Channel Message

This message contains the new RF channel number and a list of TEIs. This message is used by the MDBS RRME to force the selected M-ESs to switch to the new channel defined in the message.

In CDPD implementation, the switch channel message can be employed in the following operating scenarios:

- When a shared RF channel is being used in a channel stream and the dwell timer is set to nonzero, the MDBS will initiate a planned hop to a new channel. To make an orderly transition, the MDBS will transmit the switch channel message. In this case, since all the M-ESs using the current channel need to hop to the new channel, the message will contain ONE TEI value defined as 0.

- In a typical load balancing algorithm where periodically an RRME algorithm evaluates the load on different channels of the same cell, it may determine a new distribution of M-ESs to channels that may provide more efficient use of the RF bandwidth. In this case, the list of M-ESs may be identified and a switch channel message sent to the selected list.

Bit	8	7	6	5	4	3	2	1
Octet 1	LMEI ($=42_{10}$)							
Octet 2	Type = Switch Channels (04_{10})							
Octet 3	Reserved							
Octet 4	RF Channel Number							
Octet 5	Number of TEIs							
Octet 6...	TEI 1							
•••	TEIs 2 to N							

FIGURE 7.13 Switch channel message format. (*Source:* [2]. © CDPD Forum.)

The message parameters are shown in Figure 7.13.

7.4 KEY FUNCTIONS OF RADIO RESOURCE MANAGEMENT

7.4.1 RRMEs

The MDBS radio resource management serves as the primary entity that collects configuration information and broadcasts the relevant information as required by the RRMP protocol. The RRME in the MDBS will use the services of the physical layer, the MAC layer, and the MDLP layer to carry out its functions. These functions include RF channel selection, managing the tuned channel (including initiation of planned and forced hops), and channel congestion management.

The RRME in the M-ES will extensively use the services of the physical layer, the MAC layer, and the MDLP layer to carry out its functions. The RRME functions in the M-ES include processing of RRMP messages and selecting the best RF channel for setting up and maintaining the MDLP connection with the network.

An additional RRM function that is related to sleep management of appropriately equipped M-ESs is carried out by cooperating entities that reside in the MD-IS and in the M-ES.

7.4.2 MDBS RRME Functions

In addition to broadcasting the RRMP messages, the RRME entity resident in the MDBS assigns frequencies to traffic channels. The RRME also manages re-assignment of frequencies due to implementation-dependent conditions or when frequencies are forced out from CDPD by voice activity.

7.4.2.1   Channel Selection

One of the main functions of the MDBS RRME is to select and assign an RF channel for each channel stream configured in a cell.

A typical CDPD cell will have a frequency pool containing a subset of frequencies assigned to the associated AMPS cell. The RRME will select a dedicated channel if one is found in the RF pool. The availability of shared RF channels in the frequency pool for use by a channel stream will be determined by one of the following implementation dependent factors:

- CDPD systems can employ sniffers to detect the presence of a chosen RF channel in the transmitter output of the associated voice cell. Periodic scan of the shared RF channels in the CDPD frequency pool by the sniffer hardware can be used to update the availability status of the shared channels for CDPD use.
- Where the voice system implements a defined channel status protocol (CSP), the CDPD system can exchange CSP messages with the AMPS system and can update the shared RF channel status.

Once the basic availability of a shared RF channel for CDPD use is determined, implementation of specific algorithms can attempt to select an RF channel that has the lowest probability of being chosen for use in the non-CDPD (voice) system. In this respect, knowledge of the behavior of the selection algorithm in the voice system will be useful in choosing a matching algorithm for the CDPD system. The following possibilities are easily accommodated in a CDPD implementation:

- If the voice system uses a static list and selects the first unused RF channel scanning from top to bottom, the CDPD RF pool can be organized in the inverse order. Then, the best RF channel for CDPD use will be the first available when the RF pool is scanned from top to bottom.
- If the voice system uses a roundrobin algorithm, then dynamic reordering of the CDPD RF pool will become necessary. The reordering criterion is to move the RF channel that is released by the voice system to top of the CDPD list. The best RF channel for CDPD use will be the first available when the RF pool is scanned from top to bottom.
- If the voice system uses a top to bottom scanning for selection and the order of the list is not known, the CDPD system can dynamically create the best list for CDPD use by moving the RF channel when released by the voice system to the bottom of the CDPD list. The RF channels in the top of the list will then be the ones that have not been used by voice for the longest period of time. In addition, more sophisticated algorithms can be conceived that can use the continuous voice usage information provided by the sniffer subsystem to derive the long-term statistics on the channel allocation dynamics of the voice system.

7.4.2.2 *Directed Channel Hops (Planned Hops)*

A directed channel hop is a planned RF channel switching procedure initiated by the RRME on a particular channel stream. The reasons for initiating planned hops are configuration and implementation dependent.

In a directed channel hop, the RRME does the following:

- Selects the new RF-channel for use in the channel stream.
- Sends a switch channel message, and waits till the switch channel message is transmitted in the old RF channel. A burst of identical switch channel messages may be necessary to ensure that all the M-ESs tuned to the RF channel receive them.
- Ceases transmission on the channel stream and tunes to the new frequency within 40 ms of turning off the carrier on the old frequency.

The expectation of this procedure is that the M-ESs previously using this channel stream, remain on the channel stream after switching to the new frequency, thus avoiding the overhead of channel search by the affected M-ESs.

Several implementation-specific examples can be thought of where planned hops can be used:

- With each shared RF channel, a maximum channel time (swell time) value is defined. This timer, if configured as a positive value, specifies the maximum time the channel stream can use the RF channel once it has been tuned on a channel stream. On the expiration of this timer, the MDBS RRME should use the planned hop procedure to switch the channel stream to another RF channel. This timer may be set for the following reasons:

    1. The frequency assignment in the voice system may be time of the day–based and that information can be used within the CDPD system to derive a dwell-time parameter for associated CDPD RF channels.
    2. If some degree of co-channel or adjacent channel interference from an RF channel is expected based on the network design, and if there is a maximum allowed time of tolerance of this interference, the dwell time can be used to limit the amount of continuous use of the RF channel.

- If the CSP is used to interact with the voice system, the planned hop procedure may be used within the CDPD system when the voice system sends a channel busy message to the CDPD system indicating that the voice system is about to use a specific RF channel.

Associated with the dwell time, the shared RF channels are also configured with a layoff time. The layoff time defines the time a released RF channel after a planned hop cannot be used for CDPD, unless there are no other RF channels available for CDPD use.

In many CDPD networks deployed and providing commercial service, the dwell time for shared channels are configured as 0. This implies that if a shared channel is selected for use, it will continue to be used until forced out by voice activity in the AMPS system.

7.4.2.3   Undirected Hops (Forced Hops)

Undirected hops or forced hops are used by the MDBS RRME when time constraints prevent the MDBS from being able to carry out the planned hop procedure. When the non-CDPD activity is detected on an RF channel currently being used in the CDPD system, the requirement mandates that the CDPD should relinquish using the RF channel within 40 ms. This stringent time constraint, which includes the latency in detecting the presence of non-CDPD (primarily voice) activity by the sniffer subsystem employed in the MDBS, does not allow the MDBS sufficient time to find a new RF channel, and generate and transmit a switch channel message before ceasing transmission on the old RF channel.

Typical sequence of operations that is carried out during a forced hop is as follows:

- Non-CDPD activity is detected on an RF channel used by the CDPD system. The latency between the start of non-CDPD activity and the detection of it depends on the sniffing hardware and the sniffing methodology employed. To reduce cost in the deployment of the sniffing hardware, the worst case latency will be designed to be closer to 30–35 ms so that there is little time left to do any additional processing.
- The RRME uses the physical layer to immediately turn off the carrier to satisfy the 40-ms time constraint. If the overall time constraint allows, then an attempt can be made to minimize the loss of transmissions by suspending the forward data transfer before turning the carrier off.
- The M-ES will detect a channel exception condition primarily as loss of synchronization detected by the MAC layer and will choose a new channel stream for continuation of its activity.

Implementation of an efficient and fail-safe system for detecting non-CDPD activity is a critical requirement for a successful CDPD system. Extensive use of self-testing methodology needs to be implemented so that the system marks the shared frequencies unusable if failure is suspected in any part of the sniffing subsystem.

Even in cases where the CSP is used with the voice system, scenarios can be identified where forced-hop procedures have to be used during channel hops so that the delays in the CSP protocol messages do not violate the 40-ms limit on CDPD channel switch-off time.

7.4.2.4 Channel Congestion Management

The channel capacity flag in the channel stream identification message provides the MDBS RRME with the capability to inhibit additional M-ESs from registering on a channel stream.

The capacity flag reflects the internal determination of the CDPD system whether the specific channel is congested. The specification requires that if there is only one channel stream in a cell, the capacity flag should be 0, indicating that the channel is never congested. The strategy used in the MDBS (and possibly in the MD-IS) to manage congestion in a channel will increasingly become important as the number of traffic channels per cell increases in deployed networks to handle increasing CDPD traffic.

When multiple channel streams per cell are employed, an efficient channel congestion management strategy should take into account the following scenarios:

- Channel congestion depends on the traffic volume carried by the channel in the forward and reverse direction. Thus, to build a congestion control scheme that can adapt to changing traffic flow conditions, continuous measurement of the forward and reverse channel traffic on a per M-ES basis becomes necessary.
- In M-ES applications where only bursty traffic is used, the number of M-ESs registered on a specific channel stream may also be important.

A workable algorithm will use a suitable combination of the number of M-ESs and their throughput to derive a dynamic congestion control strategy. In addition to making the channels that are carrying excessive traffic marked as capacity full, switch channel messages can be targeted to selected M-ESs, forcing them to change to other channels.

7.4.3 M-ES RRME Functions

The RRME resident in the M-ES uses the RRMP messages broadcast by the MDBS for performing its radio management functions.

7.4.3.1 RRMP Broadcast Message Processing

Processing the received RRMP messages and maintaining the current cell and its neighbor information is one of the main activities of the M-ES RRME.

7.4.3.2 Wide Area Channel Search

When an M-ES is first powered up, the procedure the M-ES RRME executes to initially acquire channel is called a wide area channel search. The order of search for a usable RF channel is implementation-dependent and some possible algorithms are as follows:

- The most simple form will be a sequential search of all possible RF channels (800 channels of 2–5 ms/channel taking up to 3 sec), or to limit the search to carrier A or B based on the configuration of the M-ES with the knowledge of the carrier (WASI) operating in the geographic location.
- A search order based on the retained knowledge of the RF channels in cells previously acquired by the M-ES.

The following is the sequence of operations the RRME schedules during the initial acquisition of an RF channel:

1. The RRME selects the next frequency from its search list and invokes the physical layer function to measure the RSSI value of the selected RF channel.
2. If the RSSI value is not within an acceptable range determined by implementation-dependent criteria, the next channel is selected and the procedure is repeated from step 1.
3. On acceptable RSSI value, the RRME invokes the MAC layer function to synchronize on the chosen RF channel (MAC-OPEN request).
4. If the MAC layer is unable to synchronize (if a MAC-CLOSE indication is received), then the procedure again continues from step 1.
5. If the MAC layer is able to synchronize to the RF channel (i.e., the sync patterns on the forward channel are detected by the receiver), the RRME continues to measure the quality of the channel. The quality parameters are implementation-dependent and are generally based on a combination of sync error rate, block error rate, and symbol error rate.
6. The RRME then waits until it receives the channel stream identification message.

At this stage, the RRME continues to execute what is defined as a cell transfer procedure, which is also used in several other operational scenarios, and therefore is described as a separate sequence of operations in Section 7.4.3.4.

Since the wide area channel search is a fallback, all-inclusive search strategy, the M-ES will execute this as a last resort when more efficient strategies, such as adjacent cell scan, fail.

7.4.3.3 Adjacent Cell Scan

In the newer version of the CDPD specification, the M-ES RRME is required to "continuously" measure the signal quality of the current cell and neighbor cells. In these periodic evaluations, if the M-ES locates a "better server" channel, then the current channel the M-ES is expected to switch to that channel and to execute a cell transfer procedure. This scheme provides the best CDPD channel for an M-ES since the channel selection method is based on relative strengths of RSSI values.

Terrain Effects and Scan Efficiency

Terrain effects can produce adequate signal quality from a distant cell well into the *cell boundaries* of an adjacent cell. An M-ES continuation of using this channel from a distant cell can cause interference from its strong reverse channel transmissions on the CDPD or voice transmissions in other cells. The best server channel selection helps to avoid this possibility. Further, adjacent cell scan allows techniques for adjusting the measured signal quality to account for terrain, traffic conditions, and unnecessary switch-over flipflops between two cells. The specific parameters and strategies that allow the M-ES to deal with the terrain conditions and make the adjacent cell scan efficient are as follows.

Reference Channels and ERP Delta During the best channel selection, initially only the reference channel defined for the neighbor cells needs to be scanned to select the best cell. These reference channels are carried in the cell configuration broadcasts relevant to each neighbor cell. During comparison of RSSI measurements, the measured value of the RSSI level has to be adjusted by the ERP delta parameter before the value is compared those of other neighbor cells. The ERP delta value represents the difference in the RSSI values of a CDPD broadcast channel and the reference channel, which can be an AMPS control channel or a CDPD dedicated channel. The ERP delta is a configurable parameter and is broadcast along with the value of the reference channel in the cell configuration broadcast.

RSSI Bias The RSSI bias value is defined for each neighbor relationship and is broadcast in the cell configuration message. When comparing the RSSI levels of the different neighbor cells, each reference channel RSSI value is further adjusted by the neighbor's RSSI bias value before comparison. The effect of this adjustment is to effectively move the cell boundary towards or away from the specific neighbor cell depending on the sign and magnitude of the RSSI bias value. This feature can be used for load sharing between neighbors by moving the cell boundary of the higher traffic cell towards its center.

RSSI Hysteresis When there is a heavy volume of traffic along a boundary of two cells with equal power levels, best channel selection may cause unnecessary flipflops between the cells. This parameter provides a hysteresis effect during the best channel selection and avoids superfluous channel reacquisitions.

The best server channel selection, therefore, begins with a best cell selection followed by the channel selection within that cell, and subsequently channel acquisition.

Reference Channel Scan Procedure

The reference channel scan procedure, whereby the RRME in the M-ES selects the best cell for CDPD use can be described as follows:

1. For the next adjacent cell for the current cell, find the RSSI value of its reference channel, $RSSI_{RefCh}$.
2. Compute the effective RSSI value of this neighbor cell.

$$RSSI_{Eff} = RSSI_{RefCh} - ERP_{Delta} + RSSI_{Bias}$$

   Note that smaller ERP_{Delta} and larger $RSSI_{Bias}$ will make the neighbor a more attractive cell for CDPD use in the current scan.

3. Compute the $RSSI_{Eff}$ for all neighbors (repeat steps 1 and 2) of the current cell and select the best cell or cells with the highest $RSSI_{Eff}$.
4. Compare the highest neighbor to the RSSI value of the current cell ($RSSI_{Curr}$) taking into account the $RSSI_{Hysteresis}$ value to determine whether the M-ES should hop to another cell. The M-ES must proceed to reacquire the current channel if

$$RSSI_{Curr} + RSSI_{Hysteresis} > Max\ RSSI_{Eff}$$

5. If the current cell is the best cell, then the RRME:
   - Tunes to the previously used channel stream;
   - Restarts the RSSI scan timer to trigger the next adjacent cell scan;
   - Sets the internal RSSI value, $RSSI_{InternalRef}$, to the mean RSSI value of the current channel. As described in the next section, this value is used in determining another triggering condition for the adjacent cell scan, and the adjacent cell scan procedure terminates. The M-ES continues to use the same channel stream.

6. If there is more than one best candidate cell group, the ones with the same area color code are given priority over the others. The RF channels in the chosen cell are scanned one after the other until an RF channel stream is found which satisfies:
   - $RSSI_{New-Stream} > RSSI_{RefCh} - ERP_{Delta}$;
   - Implementation-dependent channel quality criteria such as block error rates are acceptable.

The RRME then executes the cell transfer procedure.

Reference Channel Scan Triggers

The best server channel selection and the adjacent cell scan are triggered by two different conditions:

1. A configurable periodic timer determines the maximum time between two adjacent cell scans. It is started during the initial power up of the RRME application, and is retriggered after each adjacent cell scan. The recommended value of this timer is 90 seconds.
2. Rate of change of RSSI value: If the RSSI average value of the current cell traverses outside the range defined by $RSSI_{InternalRef} + RSSI_{ScanDelta}$, and $RSSI_{InternalRef} - RSSI_{ScanDelta}$ for a period longer than the configured RSSI average time, an adjacent cell scan procedure is executed. In the case when the RSSI average is increasing (condition where current RSSI value exceeds $RSSI_{InternalRef} + RSSI_{ScanDelta}$), only the face neighbors are scanned.

When the adjacent cell scan is initiated by an increasing RSSI average value, the likely scenario is that the M-ES is moving towards the center of a sectored cell. As shown in Figure 7.14, the positively increasing RSSI average trigger forces the M-ES to perform a face neighbor scan so that M-ES chooses the face neighbors channel as it crosses the face boundary. Without this trigger condition, the M-ES is likely to hold on to the current channel well into the face neighbor's territory.

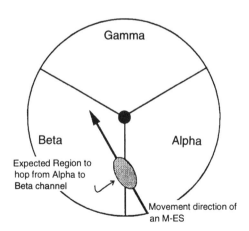

FIGURE 7.14 Increasing RSSI trigger of face neighbor cell scan.

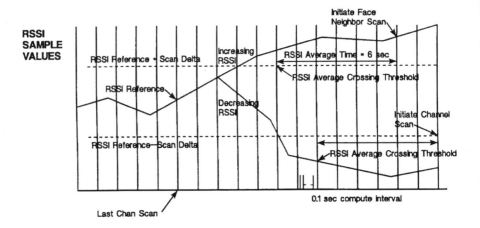

FIGURE 7.15 RSSI measurement procedure.

RSSI Measurements

Figure 7.15 illustrates the technique employed by the M-ES RRME to measure the RSSI values of the current channel and shows the conditions when adjacent cell scan procedure is triggered.

As shown in Figure 7.15:

- RSSI value is computed ten times every second.
- Only samples taken in the last 5 secs are used in computing the RSSI average.
- RSSI samples are flushed when the RF channel is changed.
- If the time elapsed is less than 6 sec, only the samples collected are used in computing the average.
- RSSI average must be outside the range for more than the RSSI average time to initiate face neighbor scan or the adjacent cell scan.
- If RSSI average value exceeds the defined range (positive rate of change), only the face neighbors are chosen for the cell scan.

BLER Measurements

Figure 7.16 illustrates the technique employed by the M-ES RRME to measure the block error rates. This procedure is carried out by the RRME as it continuously assesses the quality of the channel that it is using.

As shown in Figure 7.16:

- BLER is computed once every second.
- BLER computation is initialized when the RF channel is changed.

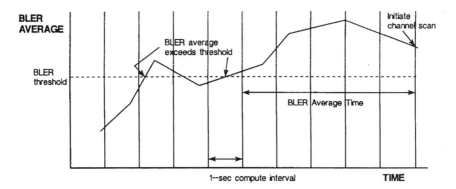

FIGURE 7.16 BLER measurement procedure.

- Only the number of intervals within the BLER average time is used in the calculation of the average BLER.
- If less time has elapsed than the configured BLER average time, BLER the averaging procedure will assume the rest of the intervals as having no errors.
- BLER average must be above the configured BLER threshold for a length of BLER average time to result in a channel exception condition and to initiate an adjacent cell scan.

Channel Quality

Channel quality is continuously monitored by the RRME, and several conditions can render the channel unusable. These channel exception conditions initiate the adjacent cell scan procedure. The following is a list of channel exception conditions:

- Measured mean block error rate exceeding configured threshold value.
- MAC layer reporting error due to loss of channel synchronization.
- MAC layer reporting error due to the reception of invalid group or area color code.
- MAC layer reporting error due to failure of maximum transmit attempts (configurable number).
- Other implementation-dependent error conditions, such as number of decode failures reported by the MDBS as decode flags in the forward channel stream or number of forward sync words in error, may also trigger the MAC layer to report channel exception error.

In a properly designed network with adequate radio coverage, cell hops triggered by channel exception conditions should be rare.

7.4.3.4 Cell Transfer Procedure

The cell transfer procedure is a sequence of actions carried out by the RRME after the reception of the channel stream identification message, subsequent to the selection of a new RF channel, for continuation of the M-ES use of the CDPD network. The following events will typically trigger the M-ES to select a new RF channel:

- Completion of a wide area channel search;
- Completion of an adjacent cell scan;
- Completion of a directed channel hop that moves the M-ES to a different channel stream.

The set of operations executed by the RRME as part of the cell transfer procedure are as follows:

1. A check is made if the SPNI, SPI, or WASI received as part of the channel stream identification message are acceptable to the M-ES. If it is unacceptable, the M-ES will select the next best channel from the list (wide area scan or adjacent cell scan) and continue the search.
2. The internally maintained value of the RSSI reference to the mean RSSI of the selected channel is set and the RSSI scan timer is started.
3. The power product and the maximum power level value are set to that of the current cell.
4. The RRME initializes for BLER and RSSI measurements, and, if it is a new cell, initializes the internal cell configuration information base.
5. If the area color code has remained unchanged, the RRME instructs the MDLP layer to resume operation. If the M-ES has a new area color code, then interarea cell transfer procedure is followed.

Interarea cell transfers require the RRME to clear the current MDLP connection, to set up a new MDLP connection, and to reregister using the mobile network registration protocol (MNRP) with the new MD-IS supporting the newly acquired channel stream.

7.4.3.5 Switch Channel Message Processing

A switch channel message is sent from the MDBS typically during a planned hop triggered by the expiration of the dwell timer or as required by the specific implementation of a load balancing scheme in the MDBS. The message contains the new RF-channel number and a list of TEIs associated with the M-ESs. If all M-ESs that are tuned to the channel stream are supposed to switch, the TEI list will contain one entry, a TEI value of 0.

On reception of a switch channel message, the receiving M-ES is expected to switch to the new frequency. If the new frequency is supported in a different traffic channel, as in the case of M-ES load balancing scheme, the carrier is likely to be already active, and the procedure of tuning to the new frequency will be straightforward. However, in the case of dwell timer expiration, the MDBS has to transmit one or more switch channel messages, turn the current carrier off after the last message has been transmitted, and tune to the new frequency. At the M-ES end, as soon as the M-ES processes the switch channel message, it will attempt to tune to the new frequency contained in the message. A fast M-ES may attempt to tune to the new frequency before the MDBS has time to tune its carrier and may fail in acquiring the intended frequency. Strict timing requirements to ensure successful channel switching are not defined in the CDPD specifications.

7.4.3.6 Power Control

The MDBS controls the power transmitted by the M-ES with the use of two parameters, power product and maximum power level, broadcasted in the channel stream identification and the cell configuration RRMP messages broadcast on the channel stream. The maximum power level value limits the transmit power of the M-ES; the power product allows the MDBS to control the transmit power of the M-ES, if necessary. A constant power product itself has the effect of reducing the transmit power of the M-ES as it moves closer to the MDBS cell site.

Effective radiated power for the M-ESs are mapped to different power levels. These ERP values correspond to the M-ES Class-I as shown in Table 7.1.

TABLE 7.1
Power Levels of a Class-I M-ES

M-ES Maximum Power Level	Maximum ERP (dBW)
0	6
1	2
2	−2
3	−6
4	−10
5	−14
6	−18
7	−22
8	−22

TABLE 7.1 (continued)

M-ES Maximum Power Level	Maximum ERP (dBW)
9	
	–22
10	–22

The MDBS, on determining the power level at which it would like to receive transmissions from the M-ES, will send a power product value to the M-ES. In simple terms, the power product defines the following:

$$\text{Power product } [dBW^2] = \text{M-ES transmit power } [dBW]$$
$$+ \text{ mean received signal strength of the forward channel at the M-ES } [dBW]$$

As we can see, if the MDBS is transmitting at a constant power, the closer the M-ES is to the cell, the greater will be the received signal strength of the forward stream. The above rule will force the M-ES to reduce its transmit power to maintain the defined power product value.

To allow easy implementation (to avoid dealing with negative numbers), the power product value that is transmitted is computed by how much it is in excess of –143 dBW. That is,

$$\text{Power product}_{Transmitted} = \text{power product}_{Actual} - (-143 \text{ dBW})$$

Therefore, the M-ES uses the following rule to determine its transmitter power:

$$\text{M-ES transmit power } [dBW] = \text{power product}_{Transmitted} - 143 - \text{RSSI of fwd channel at M-ES}$$

The configured power product value may be adjusted seasonally, but otherwise no dynamic power control by the MDBS is deemed necessary.

7.4.3.7 Sleep Management

Sleep management procedures are defined in the CDPD specifications for implementing power conservation strategies for M-ESs. For implementation of the sleep-mode procedures, the RRME on the network side is made a subcomponent of the MD-IS software that has access to the link states managed by the MDLP layer in the MD-IS.

The overall concept is to allow the M-ES to transit into a power conservation state if there has been no messages to transmit on the reverse channel for a configured amount of time. The M-ES exits the sleep mode if it receives application frames for transmission in the reverse direction. If the network side has frames to transmit to an M-ES in the forward direction, the MD-IS based RRME includes the sleep-mode M-ES's TEI value in its TEI notification broadcasts, as shown in Figure 7.17 and explained below:

- The M-ES maintains a sleep timer (idle timer) defined as T-203, defined during the MDLP parameter negotiation activity that takes place during TEI assignment.
- The same timer is maintained in the MD-IS for each M-ES that has an MDLP session established with the network.
- The MD-IS, in addition, maintains a TEI-notification timer, T-204, and TEI notification retries, N-204, for each of the channel stream. These values are configurable.
- The MD-IS broadcasts TEI notification frames every T-204 seconds. The frame contains the TEI values of the M-ESs that have forward directional frames pending in the MD-IS. Even if there are no M-ESs using the channel stream that are in sleep mode, the notification messages will still be sent. This allows all M-ESs to synchronize to the T-204 timer of that channel stream.
- The T-203 timers are started or restarted under the following:
  1. The M-ES triggers T-203 on the completion of transmission of an MDLP frame in the reverse direction.
  2. The MD-IS starts the T-203 timer for a particular M-ES when the MD-IS completes the reception of a reverse frame from that M-ES.
- M-ESs in sleep mode, should wake up during the TEI notification message broadcast time to receive the message. Once the notification message is received, each M-ES can ascertain whether its TEI value is included in that message, indicating that there are pending messages in the MD-IS for this M-ES. In this case, the M-ES sends an RR frame to force the MD-IS to mark this M-ES as not sleeping.

The RF channel the M-ES was using may not be available for use when the M-ES wakes for the following reasons:

- Planned hop or forced hop may have removed the last used RF channel from the channel stream.
- M-ES may have relocated to a new physical location so that the last channel stream that it was previously using may no longer have adequate signal strength for the M-ES to receive the forward directional transmissions.

As long as the M-ES can synchronize to the forward channel, it will attempt to use the last used cell. If an implementation-dependent RSSI level criteria is not met

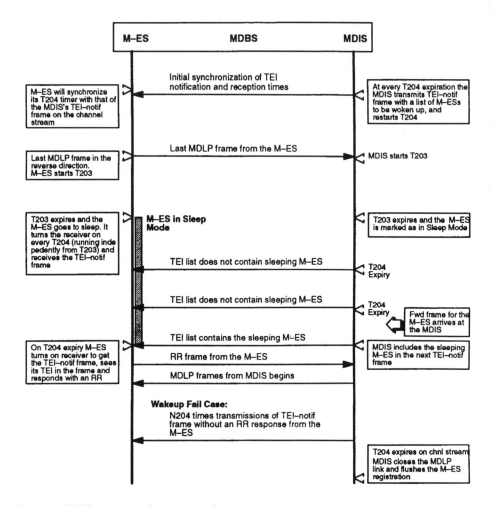

FIGURE 7.17 Operational scenarios in sleep management.

for the last used channel, then the M-ES may attempt to execute the channel acquisition procedure to find a better channel on the same cell. Only if a channel was not acquired when the M-ES went to sleep (received the sleep indication notification), the M-ES will resume the adjacent cell scan or the wide area channel search to find a new best server on waking up.

7.5 RADIO FREQUENCY COVERAGE

Radiofrequency footprints are key to providing acceptable radio coverage to the CDPD system. As a dependent service on AMPS, the flexibility in being able to configure and tune a CDPD network is therefore limited by the constraints imposed by the AMPS network. Some dependent elements are as follows:

- CDPD base stations are co-located with AMPS base-stations, generally on a one-to-one basis, and therefore cell/network boundaries are expected to be the same.
- The transmitter output power for the AMPS system is determined after careful radioengineering studies and will define a cell-boundary. In addition, the frequency reuse pattern to avoid co-channel interference would limit the maximum power. The CDPD system must adhere to this limit; violating this rule will cause co-channel interference on an AMPS frequency by the CDPD transmission.
- The definition of cell "boundaries" depends on the RF foot print as well as the handoff criteria and parameters used by the AMPS and the CDPD systems.
- The power output specifications of the M-ES and the M-ES/MDBS receiver sensitivities must have comparable values. Compliance of the MDBSs and M-ESs are a prerequisite for interoperability of M-ESs and carrier networks.

7.5.1 Cell Boundaries in AMPS and CDPD

Mobile receiver sensitivities have a key impact in the "virtual" cell boundary demarcation in a CDPD network. In this respect there are important differences in the operation of the AMPS network and the CDPD network:

- In the AMPS network, a mobile is assigned an RF channel for the duration of a call. Channel handoff is mainly determined by the cell site. The cell has access to control channels and special scanning receivers are used to determine the best channel.
- In the AMPS system, both mobile-initiated handoffs and base-station-initiated handoffs are possible because of the exclusive allocation of the forward and reverse channels to a call. In CDPD architecture, the reverse channel is a multiple access channel and the MDBS initiating a handoff is not possible. The scheme in CDPD, therefore, is to make the M-ES responsible for initiating and executing the handoff.

Effort has to be made during commissioning to ensure that the CDPD cell boundaries nearly coincide with the AMPS cell boundaries. If cell dragging is present to a significant level in CDPD, this would potentially result in strong M-ES transmission, producing co-channel interference in cells in a different group (cluster).

Further, the general expectation of the CDPD network operating personnel is that once the transmit power levels of the AMPS and CDPD systems are adjusted, the cell boundaries have to be identical. That is, even in fringe areas of RF coverage, if a cellular phone can successfully establish a call, a CDPD M-ES is expected to register with the network and to successfully transfer data. Although this assumption is valid in properly engineered networks, in sparsely deployed sites, the AMPS system may perform better than the CDPD system. The main reason for this is that the AMPS handsets are expected to work in lower received RSSI levels than the M-ES. A more detail account of the RF-related specification issues are given in the following sections.

7.5.2 Comparison of Receiver Sensitivities

For users of CDPD and other wireless technologies, the following information related to receiver sensitivity requirement may be useful. The figures are only approximate values:

- The noise temperature representing the radio environment translated to power levels: –129 dBm.
- The noise figure of the LNA (low noise amp): 6 dB.
- Therefore, noise at the receiver: –123 dBm.
- The following is assumed to be the minimum carrier to noise ratio (CNR) required for successful detection:
  - Fixed receiver: 6–7 dB;
  - Mobile (fading with diversity): 12 dB;
  - Mobile (fading, no diversity): 17 dB.
- The above figures provide the following approximate values:
  - Minimum receiver sensitivity (fixed): –116 dBm;
  - Minimum receiver sensitivity (fading): –106 dBm.

The values provided in the different specifications [2–4] can now be interpreted based on the above rule of thumb calculations.

7.5.2.1 Analog Voice

In the analog voice, the sensitivity requirement is satisfied by 12 dB or higher SINAD being measured when operating with the antenna in an RF field strength equivalent to the receiver input as shown in Table 7.2.

TABLE 7.2
Receiver Sensitivity for Analog Voice

Receiver Input (dBm)	RF Field Strength at Antenna
–116	1.3×10^{-14} mW/cm2 7 μV/m
–113	2.6×10^{-14} mW/cm2 10 μV/m

The SINAD of a baseband output signal is defined as the ratio of the total output power to the power of the noise and distortion expressed as

$$\text{SINAD} = (\text{signal} + \text{noise} + \text{distortion})/(\text{noise} + \text{distortion})$$

7.5.2.2 IS-54 Digital Voice

For the digital case, the specification enforces a limit on the bit error rates (BERs) for different signal levels measured for simulated flat Rayleigh fading and static (no fading) conditions. The actual performance measured in BER shall not be worse than that shown in Table 7.3.

TABLE 7.3
Receiver Sensitivity for Digital Voice

Simulated Vehicle Speed	RF Level (dBm)	BER
Static	−110	3%
Faded 8 km/h	−103	3%
Faded 100 km/h	−103	3%

7.5.2.3 CDPD

Sensitivity specifications for CDPD are based on block error rates (BLER) defined as follows:

$$\text{BLER} = (\text{blocks sent} - \text{correctable blocks received})/\text{total blocks sent}$$

A correctable block is a received Reed-Solomon block that either contains no errors in the 378-bit block or that contains up to and including the maximum number of symbol errors that can be corrected by the (63,47) code word. Reed-Solomon decoding will typically correct up to seven symbol errors.

All RSSI measurements indicate the lowest signal level required to provide 5% BER (see Table 7.4).

TABLE 7.4
Receiver Sensitivity for CDPD

Simulated Vehicle Speed	RF Level (dBm)	BLER
Static AWGN	−113	5%

TABLE 7.4 (continued)

Simulated Vehicle Speed	RF Level (dBm)	BLER
Faded 8 km/h	−100	5%
Faded 50 km/h	−101	5%
Faded 100 km/h	−102	5%

M-ES radio modems that occupy a total volume of 80 cubic centimeters or less are considered small form factor devices. For these devices, the specifications allow reduced sensitivity requirements, defined in Table 7.5.

TABLE 7.5
Receiver Sensitivity for Small Form Factor Modems

Simulated Vehicle Speed	RF Level (dBm)	BLER
Static AWGN	−111	5%
Faded 8 km/h	−98	5%
Faded 50 km/h	−100	5%
Faded 100 km/h	−101	5%

For the AMPS, the receiver sensitivity for the fading environment is not specified in the standards. This may be due to the fact that the analog voice quality below the threshold deteriorates gradually as opposed to the sharp deterioration in performance seen in the digital (voice and CDPD) system as the RSSI levels drop below the minimum level specified in fading environments.

The sensitivity figures for digital voice and CDPD are close to each other, differing by about 3 dB or less. The difference may be attributable to the measurement criteria for voice as opposed to data transfer. For voice, BERs will determine the frame error rates out of the vocoder, which will determine whether one has "acceptable" voice quality. A 3% BER is assumed to be the measurement point and any higher BER is expected to generate higher frame error rates that render voice quality unacceptable. For CDPD, 5% BLER is assumed to be the measurement point for the receiver sensitivity. In the field, however, throughput performances that are not significantly different have been seen for BLERs up to 10%. If this behavior can be shown to be consistent, the CDPD minimum performance RSSI thresholds can be lowered somewhat (3 dB) and the receiver performances can be assumed to have same RSSI thresholds for voice and CDPD.

7.5.3 Power Balance

In practice, it is expedient to balance the radio channel power levels in the forward and reverse directions to comparable levels. Various components that amplify and attenuate the radio transmissions in both directions have to be identified and considered in the analysis.

On the M-ES side, the main constraints are the transmit power and the gain of the antenna used in the mobile system. The M-ESs are categorized into classes based on the power range of their transmit channel, as listed in Table 7.6.

TABLE 7.6
Transmit Power Ranges for the Different M-ES Classes

Class	Maximum (dBW)	Minimum (dBW)
I	6	–22
II	2	–22
III	–2	–22
IV	–2	–34

The antenna gains of the M-ESs are relatively small because of the size constraints and are in the range 0–3 dB compared to the 6–12 dB of those in the base stations.

The effective radiated power (ERP) of the base station in the forward direction, which is 8–12 dB higher than that in the reverse direction, is required to compensate for the deficient receiver circuitry performance of the mobile equipment. The noise factor of the portable/mobile receiver is worse by 1–3 dB and the receiver CNR required in the portable/mobile is about 4 dB more than that in the MDBS.

References

[1] Balston, D. M., and R. C. V. Macario, *Cellular Radio Systems*, Norwood, MA: Artech House, 1993.

[2] CDPD: System Specification Book 2, Parts 300–630.

[3] IS-55: Recommended Minimum Performance Standards for 800 MHz Dual Mode Mobile Stations.

[4] IS-56: Recommended Minimum Performance Standards for 800 MHz Base Stations supporting Dual-Mode Mobile Stations.

CHAPTER 8
▼▼▼

MOBILITY OF M-ESs

8.1 INTRODUCTION

This chapter examines the mobility aspects of the M-ESs. Specifically, it details how the functionality of the CDPD protocol is structured to handle the addressing and routing of information to end systems that are mobile. The chapter also examines the concepts employed in a similar existing protocol, the mobile-IP, and discusses the differences between both approaches.

The M-ESs are assigned fixed addresses (e.g., IP or CLNP), but they can change their subnetwork point of attachment (SNPA)—the subnetwork to which they are connected to defined by the cell id and the channel stream identifier—as they move. Traditional protocols employed in routing and connectivity cannot be employed to determine the location of a particular M-ES from its address. CDPD-specific techniques that draw from similar work in the areas of aeronautical communications and the Internet have been specified and implemented to provide efficient dynamic end-to-end path for moving end systems.

Two protocols are defined that support the network layer mobility in the CDPD networks: the mobile network location protocol (MNLP) and the mobile network registration protocol (MNRP).

These two protocols, based on the concept of a home MD-IS for each or group of M-ESs, provide the basis for mobility management in CDPD nets.

In the current stage of CDPD implementation, the main subnetwork profile supported by the M-ESs and the F-ESs are IP-based. Provision of an SNDCP layer within the CDPD communication architecture allows the hooks necessary to add CLNP-based systems as the CDPD evolves. However, the inter-MD-IS communication has migrated to CLNP, as the routers (intermediate systems) supporting the CLNP protocol and the associated ES-IS, IS-IS level-1, and level-2 routing protocols have become available. This chapter therefore has a brief introduction to the routing framework in the OSI networks.

8.2   MOBILITY LEVELS AND ROUTING OF NPDUs

In this section, we review the different levels of mobility and the different CDPD infrastructure components, services and protocols that facilitate management of mobility. We can group the CDPD infrastructure components broadly into two categories: the CDPD components and the existing OSI/Internet components.

The overall concept is to isolate the mobility aspects of the M-ESs from the routing components of the existing IP/OSI networks, and to confine the mobility awareness routing within the CDPD network components (MDBS/MD-IS). The CDPD-specific components are shown in Figure 8.1 and described next.

8.2.1   CDPD Entities

8.2.1.1   Cell

A cell is defined by the geographic area covered by a single MDBS, or in the case of sectorized cells, the coverage area of the sector. A cell can support multiple channel

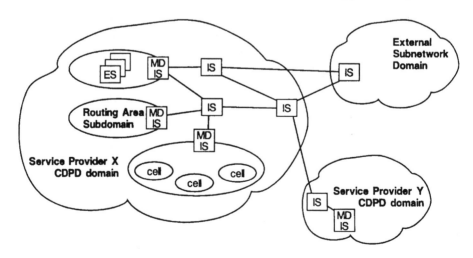

FIGURE 8.1   Cells, areas, and domains. (*Source:* [1].)

streams for the purpose of sharing the load from M-ESs distributed within its coverage area. The lowest level of "mobility" is concerned with the CDPD network supporting the transfer of an M-ES from one channel stream to another within the same cell. This may be the result of the following:

- The MDBS instructing an M-ES to move to another channel stream as part of the MDBS's load-balancing strategy.
- When a traffic channel is forced off a shared RF channel because of voice activity, the traffic channel has to request the RRME to allocate a new RF channel. Two scenarios are possible:

  1. Even if a new RF channel is allocated and the new RF channel is tuned with minimum delay, the M-ESs that were using the old frequency may not end up on the same traffic channel. The choice of a new channel stream depends on implementation-dependent frequency scanning algorithms in the M-ESs and also the relative time delays of activities in the MDBS and the M-ESs. For example, the MDBS will take a finite amount of time from the time the carrier is switched off to the tuning of a new frequency (typically 10 to 30 ms or more). The M-ES has to determine the loss of sync condition, and start scanning the frequencies in the pool of that cell in some implementation-dependent sequence. Further, the tuning and RSSI sensing time for an RF channel, when the M-ES scans the frequencies depends on the hardware and can vary between 2 ms to 10 ms or more.
  2. The RRME may not be able to deliver a new frequency if temporarily the shared RF channels are all busy with voice activity. The M-ESs that used the old RF channel will end up choosing another channel stream on the same cell during its channel search activity.

In the perspective of mobility management, an intracell transfer (one stream to another in the same cell) or an intercell transfer among cells controlled by the same MD-IS are conceptually similar. This is due to the fact that the network side of the MDLP termination point resides in the MD-IS, making the stream within a cell or different cells of little consequence to the MD-IS processing of the MDLP layer. However, note that the RRME aspects of an intercell transfer are markedly different from an intracell transfer.

8.2.1.2 Area

The combined geographic coverage of all MDBSs under the control of a single MD-IS is defined as a routing area. The *area* concept is important in that mobility within a routing area uses different mechanisms than mobility between different routing areas. The MNLP provides the core facility to manage the location information of a moving

M-ES. A routing area may cover multiple cellular geographic service areas (CGSAs) or it may be only a part of a single CGSA.

The concept of home area and serving area provides the framework for the implementation of the MNLP protocol. Each M-ES is logically a member of a fixed home area. The home area provides the CDPD system the mobility-independent anchor to the network ISs and other ESs that are not aware of the mobility of the M-ESs.

8.2.1.3   Domain

A CDPD domain is defined by the set of MD-ISs operated and administered by a single CDPD provider. Mobility between CDPD service provider domains is subject to service agreements between those providers, but in principle is no different from mobility between routing areas.

8.2.2   ISO/IP Routing Framework

Overall CDPD network architecture will be likely to contain an aggregate entity of small backbones for each service provider, connected by IP or OSI nets, with interconnection between providers implemented with cooperative agreement. Figure 8.2 illustrates the interconnection of the different network entities that will participate in providing service to a visiting M-ES in a different service provider's domain.

These network entities involved in routing within the CDPD infrastructure fall into the following categories:

- Boundary intermediate systems (BIS) for interservice provider routing;
- Regular intermediate systems (ISs), which will serve as relay stations within a routing domain to provide intraservice provider routing;
- MD-ISs, which provide the mobility management functionality, are not ISs in true sense, but need routing protocols to interface with a service provider's IS. The MD-IS to IS interface involves:
  1. Routing information exchange for inter-MDIS communication using CLNP protocol;
  2. Routing information exchange for supporting IP-based data traffic between mobile and fixed-end systems.
- Not represented as a specific block in the figure but resident as distributed function in the CDPD network are the MNRP/MNLP protocols, which provide the routing/reachability information for the M-ESs.

The recommended routing and addressing plan for the CDPD network involves either the Internet protocol (IP) or the ISO 8473 [2] connectionless network protocol (CLNP) to be used as the network layer in the infrastructure components and the end

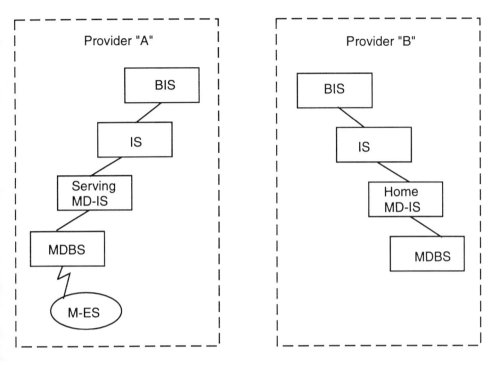

FIGURE 8.2 Network entities providing service to a visiting M-ES.

systems. Although IP-based Internet is the predominant protocol currently being used for application to application communications, CLNP-based communication is the only type specified for the inter-MDIS communications. The protocol selections for use in the different CDPD network entities are listed in Table 8.1.

TABLE 8.1
Protocol Support in CDPD Network Entities

Network Entities	Protocols
Application Service F-ES	TCP, IP, ICMP
Support Service F-ES	TP4, CLNP, ES-IS
Management F-ES	UDP, IP, ICMP
Internet F-ES	IP, ICMP
OSI net F-ES	CLNP, ES-IS
MDBS	CLTP, CLNP, ES-IS

TABLE 8.1 (continued)

Network Entities	Protocols
MD-IS	OSPF, IP, ICMP, CLNP, IS-IS, MNLP
IS	OSPF, IP, ICMP, CLNP, ES-IS, IS-IS
BIS	OSPF, IP, ICMP, CLNP, ES-IS, IS-IS, IDRP, BGP-4

Within the OSI terminology, level-1 routing works within an *area*, level-2 routing connects areas and works within a routing domain, and level-3 routing (interdomain routing) connects routing domains.

A brief description of the routing entities that are not mobile-aware, but form the overall infrastructure of the CDPD routing framework, are given in the following sections.

8.2.2.1   Interservice Provider Routing

The OSI addressing format provides a convenient scheme for implementing hierarchical routing, and at the highest level of interest to CDPD are routing domains. Interoperation with systems in other administrative domains (different service provider components) is conducted in a mutually suspicious manner. Protocols that consider the security issues and administrative policies have to be implemented for interchange of packets between two *administrative* domains. These requirements have given rise to the routers called boundary intermediate systems (BIS), which are still being developed [3,4].

IDRP for Interdomain Routing in OSI Nets

The interdomain routing protocol (IDRP), also known as ISO 10747 [5], is being developed to accomplish comprehensive BIS functions. It is expected that the CDPD network implementations used for IDRP will support route aggregation as well as support for generation and propagation of the source- and destination-specific quality of service (QOS) and priority attributes.

BGP for Interdomain Routing in IP Nets

The boundary gateway protocol (BGP) is the IP equivalent of IDRP. In fact, the specification of IDRP is based on BGP, which has superseded the more primitive exterior gateway protocol (EGP). Use of classless interdomain routing (CIDR) [6] as a means to implement aggregation has been advocated by the CDPD forum to be supported by the BGP-4 deployment.

8.2.2.2 *Intraservice Provider Routing (Interarea)*

The interarea part of the intraservice provider routing involves the inter-MDIS communications. For the sites that have full OSI implementation, the IS-IS protocol [7] is the recommended choice. For Internet routing, the open shortest path first (OSPF) protocol has been selected.

IS-IS Level-2 Routing for Interarea Routing

The ISO-10589 IS-IS protocol is a link state routing algorithm that was adapted from the DECnet protocol. Router ports that interconnect different CDPD areas (MD-ISs) must have the IS-IS level-2 routing enabled.

OSPF Level-2 Routing

OSPF provides dynamic recovery from topological changes in the network. It also provides equal cost multipath routing by maintaining several routes to the same location. The global CDPD requirement requires the implementation of type of service (TOS) based routing in the OSPF protocol.

8.2.2.3 *MD-IS to IS Protocols (Intra-Area Routing)*

Intra-area routing is essentially performed by the MD-ISs and as such they will fall under the category of level-1 routers. However, only the interface between the IS and the MD-IS will be enabled for level-1 routing. Internally, the mobile-aware MD-ISs operate MNRP and MNLP protocols to route information to the M-ESs.

The selected protocols for the OSI routing are IS-IS level-1; implementation simplicity in the MD-ISs may also be attained by using specific ES-IS packets to inform the ISs of the reachability of the different M-ES subnets. For IP support, an appropriately configured OSPF protocol is the selected choice.

8.2.2.4 *Fixed-End Systems Routing Provisions*

Fixed-end systems that support applications running over CLNP need to have the ES-IS protocol [8] implemented to provide the reachability information to the connected level-1 CLNP routers. The equivalent protocol for the F-ESs in the Internet domain is ICMP.

8.3 MOBILITY MANAGEMENT ELEMENTS IN M-ESs

Each M-ES is identified by one or more unique network entity identifiers (the NSAP address less the selector octet), and each NEI has a single home domain (home MD-IS). The M-ES registers its NEI with the CDPD network using the mobile network registration protocol (MNRP) and the M-ES will be considered a transitory member of a current serving area; it has a serving MD-IS.

The cell id and the channel stream identification (CSI) pair is referred to as the subnetwork point of attachment (SNPA) address of the M-ES. The unique NEIs and the current SNPAs are used by the CDPD network to route messages to the M-ES.

8.4 MOBILITY MANAGEMENT ELEMENTS IN MD-ISs

MD-ISs are the only network relay/routing elements that have knowledge of the mobility of the M-ESs and execute the MNLP to exchange location information. The MD-IS's mobility management responsibilities can be broadly grouped under two functions.

8.4.1 Mobile Home Function (MHF)

Every M-ES will be configured as a member of a home area; that is, it will have a home MD-IS. The home area will serve as the anchor or mobility-independent routing destination to the other network routing elements and end systems. The MHF is responsible for performing the following functions:

- *Location directory service.* This directory maintains an information base of the current serving area for each of its homed M-ESs.
- *Redirection and forwarding service.* The forward direction information to an M-ES first arrives at the home MD-IS by traditional routing means. The MHF encapsulates these packets and forwards them to the serving MD-IS, which in turn routes them to the M-ES.

8.4.2 Mobile Serving Function (MSF)

The MSF is primarily responsible for routing all data packets for visiting M-ESs, and performs the following functions:

- *Registration directory service.* This directory maintains the information base of the M-ESs currently registered in the serving area.
- *Readdress service.* This service decapsulates the data packets forwarded by the MHF function and routes them to the destination M-Es by sending them to the appropriate cell and the channel stream (SNPA of the M-ES).

8.5 CELL TRANSFER

When radio resource management was reviewed in Chapter 7, different events that trigger the MDBS to change channels (channel hops) and the criteria and procedures for an M-ES to perform cell transfers were detailed. The M-ES, in addition to be able to acquire the best air channel for communicating with the MDBS, also has to participate in providing a mechanism for the CDPD network to implement adequate routing mechanisms. With reference to the M-ES, its ability to move manifests in the following events that have impact on the routing within the CDPD network. When an M-ES chooses a new RF channel, the new and the old RF channels

1. May be from the same cell (same cell id) and the same channel stream (same CSI), which is known as *channel hop*;
2. May be from the same cell (cell id) but from different channel streams (CSI), known as *intracell channel stream transfer*;
3. May have the same area color code (same MD-IS) but from a different cell (and therefore different channel stream), or *intra-area cell transfer*;
4. May be from cells supported by different MD-ISs (different area color code), known as *interarea cell transfer*.

Case 1 refers to a channel hop by the MDBS and the fact that RF channels (frequencies) are different on the same port on the MDBS (same channel stream identifier (CSI)) has no impact to the mobility management. The reachability and routing parameters in the MD-IS and elsewhere remain unaffected. The MD-IS that terminates the MDLP connection on the network side can be unaware of this change in frequency.

Cases 2 and 3 are similar with respect to mobility management, since in terms of routing, the MD-IS will have the SNPA of an M-ES that is defined as the following pair: <<cell-id, Channel Stream ID>>; in case 2, only the channel stream id has changed and in case 3 both have changed.

Case 4 is the more involved case, where inter-MD-IS communication between the serving (both old and new) MD-IS and the home MD-IS of the M-ES is required to maintain the routing/reachability information of the M-ES.

8.5.1 Intra-Area Cell Transfer

In CDPD networks, intra-area cell transfer—transfer between cells that are controlled by the same MD-IS—will be the dominant cell transfer mechanism.

The peer MDLP protocol layers reside in the M-ES and the MD-IS. The intracell transfer mechanism is designed not to affect the MDLP links. The MDBS serves strictly as the MAC bridge or MDLP relay between the M-ES and the MD-IS. During an intra-area cell transfer, the association between an established MDLP link

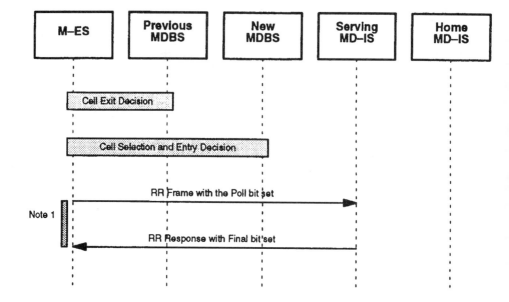

FIGURE 8.3   Intra-area cell transfer.

and the physical medium defined by the <<cell, channel stream pair>> is updated in the MD-IS and the MDLP processing will resume from the point at which it was interrupted.

The mechanism by which the new SNPA is made known to the MD-IS is shown in Figure 8.3. Once the M-ES chooses the new channel stream, it is the responsibility of the M-ES to ensure that the MD-IS receives at least one reverse direction MDLP frame. This reverse frame will be tagged by the MDBS with the <<cell-id, channel stream id>> information before it is relayed to the MD-IS. The MD-IS depends on this reverse frame to determine the new location of the M-ES and will update this SNPA information in the registration directory. Until the first reverse frame reaches the MD-IS, any forward directional frames, up to the MDLP transmit window size, can still be sent to the old cell/channel-stream and lost in the old channel stream.

Therefore, M-ES sending an RR frame with a poll bit set is normally implemented to ensure that the MD-IS receives a reverse frame after an intra-area cell transfer. The poll bit procedure ensures that the RR is retried until a response from the MD-IS is received, thus handling the cases where the RR frame is lost in the air channel.

Since the recovery of the end-to-end session is carried out at the MDLP level, the cell transfer is transparent to the layers above the MDLP (IP, TCP, or the application).

8.5.2 Interarea Cell Transfer

During an interarea cell transfer, as determined by the area color code (and the SPNI value in case the networks at a boundary are controlled by two different service providers), the M-ES establishes a new link with the newly serving MD-IS. As part of the reregistration procedure, the M-ES provides its NEI (the 19 bytes of the NSAP address) to the MD-IS.

The serving MD-IS updates its registration directory with the M-ES information in the End System Hello (ESH) packet and notifies the location update service of the home MD-IS of the redirection information using the Redirect Request (RDR) packet. Confirmation and validity of these actions are returned to the new serving MD-IS via the Redirect Confirm (RDC) packet and to the M-ES via the MD-IS Confirm (ISC) packet. The home MD-IS also sends a Redirect Flush (RDF) packet to the previous serving MD-IS enabling it to remove the entry for the M-ES from its registration directory and also to terminate the MDLP session. The interarea transfer sequences are illustrated in Figure 8.4 and detailed as part of the MNRP and MNLP sections later in the chapter.

MDLP frames and higher level packets up to the TCP level (the first connection-oriented layer above the network layer) will have been lost during the interarea cell transfer when a new MDLP link is established. The recovery from data loss will be the responsibility of the TCP (or TP4) layer.

8.5.3 Routing of Data Packets

When an M-ES originates a network protocol data unit (NPDU), it reaches the serving MD-IS. The MD-IS has access to a traditional router (intermediate system), which will have the routing information to the destination end system within the domain of the same service provider or across to the boundary routers to the destination in other domains.

The home MD-IS will be advertising the reachability of all M-ES (subnets) that are homed on it using traditional routing protocols (OSPF/RIP for IP or ISO 9542 ES-IS, or IS-IS level-1 for CLNP) to the traditional router accessible. Therefore, NPDUs destined for an M-ES, will be routed by traditional means to the home MD-IS. The home MD-IS's location directory is then used by the redirection service to relay the packet (by encapsulating it for the underlying CLNP protocol) to the serving MD-IS, which will then transmit the packet to the SNPA (cell, channel stream) address where the M-ES is located. The mechanism is illustrated in Figure 8.5.

It is seen that the NPDUs destined for the M-ES follow a different route than those generated by the M-ES. This is consistent with the conventional connectionless

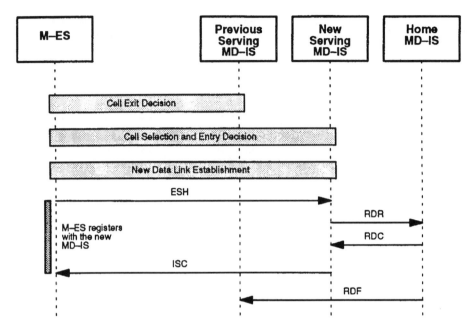

FIGURE 8.4   Interarea cell transfer.

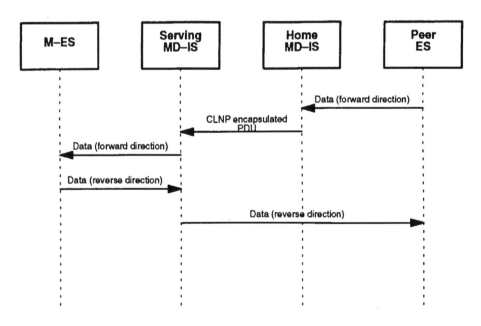

FIGURE 8.5   Routing of data packets for visiting M-ESs.

network protocol and isolates the mobility-related routing to the CDPD MD-IS elements and away from the intermediate systems that are part of the traditional terrestrial networks.

8.6 MOBILITY OF MULTICAST M-ESs

The CDPD system provides a facility to define an NSAP address that may have multiple points of presence (represent multiple M-ESs). Each multicast group member is assigned a unique group member identifier (GMID) and this information is conveyed to the MD-IS by the M-ES during registration. Also, an M-ES can be a member of more than one multicast group, each group defined by a unique multicast-NSAP address.

It needs to be observed that the multicast NEI address is known as such only to the CDPD components (MD-IS), and the off-the-shelf components that make up the terrestrial network (ISs, ESs) are not aware of the multicast nature of the address. Further, the multicast concept is unrelated to the Class-D address in the IP address domain.

The service provided by the CDPD network to the multicast group is only a one-way service (i.e., only the forward directional packets are delivered to the members of the multicast group). This imposes a restriction on the higher level applications in that only a connectionless-mode transport protocol (UDP and not TCP) can be used to transmit datagrams to the multicast group.

8.6.1 Registration and Location Update

As with unicast NEI, the multicast NEI will also have an assigned home MDIS. The M-ES registration and location update services use the unique GMID associated with a multicast NEI to distinguish the multiple points of presence of the group members.

Location directory entry for a multicast NEI in the home area managed by the mobile home function will be extended to record all different serving areas (serving MD-IS) where at least one group member is registered. The registration directory in the serving areas managed by the mobile serving function (MSF) is extended to record all SNPAs (cell/channel stream id) that are being used by at least one group member of the multicast group.

8.6.2 Redirection and Forwarding

End systems using connectionless protocols above the network layer (UDP/IP or connectionless-TP4/CLNP) can forward (broadcast) information to members of a multicast group.

The multicast redirection service, an extended MHF service in the home area, will replicate and encapsulate the original NPDU and forward it to each of the for-

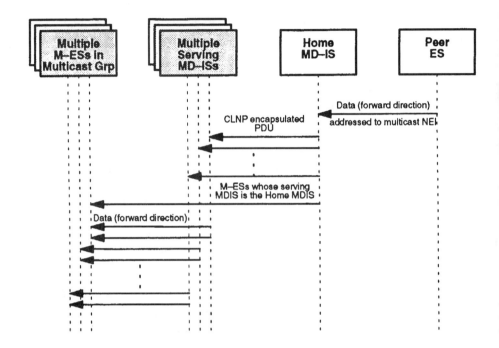

The Home MD–IS sends replicated packets to each Serving MD–IS that has at least one Multicast Group Member visiting. These MD–ISs will again replicate the data frame and send one in each of the channel streams used by members of the Multicast Group.

FIGURE 8.6 Forwarding data to multicast NEI.

warding addresses found in the location directory. When the multicast readdress server in the serving MD-IS(s) receives the encapsulated NPDU, it obtains the original NPDU by decapsulating the received packet. It queries the registration directory to obtain the different SNPA entries that indicate the cell/channel stream information to reach the members of the group that are registered with this MD-IS, replicates the NPDU, and forwards one to each channel stream that has a group member.

Figure 8.6 shows the replication, redirection, and forwarding involved in handling a data packet addressed to a multicast NEI.

8.7 MOBILE NETWORK LOCATION PROTOCOL (MNLP)

The mobile network location protocol provides for the exchange of location and redirection information between the home MD-IS of an M-ES and its currently serving MD-IS. This connectionless protocol is designed to operate over the CLNP (ISO 8473) protocol.

8.7.1 MNLP Overview

The MNLP protocol provides the main mechanism in the CDPD network to distribute the location information of an M-ES as it moves within the CDPD network. The main services rendered by the MNLP can be summarized as follows:

- *Location update function.* This is used to notify the MD-ISs of the current location of the NEIs associated with the M-ESs.
- *Forwarding service function.* The home MDIS uses this protocol to inform the serving MDIS of its willingness to provide data forwarding services for an M-ES at its current location (for the packets in the forward direction).

To provide these services, the MDIS contains mobile home function (MHF) and mobile serving function (MSF) software entities. The main functions of these two services are illustrated in Figure 8.7 and summarized next.

The MHF maintains a location directory where the current forwarding address of an M-ES homed on this MDIS will be stored and updated. The redirection service function of the MHF uses this directory to forward packets destined to an M-ES visiting a different area.

The MSF maintains a registration directory of all the M-ESs currently serving in its area. The readdress server service of the MSF receives redirected forward direction packets to M-ESs currently serving in its area, decapsulates them, and relays them to the destination M-ESs.

8.7.2 Addressing Scheme for MNLP Protocol

The two different services provided by the MNLP protocol require different network service access points to operate using the underlying network service of the CLNP. The designated NSAP selector values are shown in Figure 8.8.

It has to be noted that the location update service provided by the MNLP portions of the MHF/MSF entities is accomplished via defined packet formats based on the services provided by and restricted only to CLNP protocol. However, the redirection/readdress services part of the MHF/MSF entities have to provide forwarding services for both IP- and CLNP-type packets depending on the M-ES session type. These packets need to be encapsulated by CLNP headers for the MHF/MSF services to use the underlying CLNP protocol services.

8.7.3 Encapsulation Formats

Since only a single method of forwarding using the CLNP protocol is provided, it necessitates a scheme to identify the different network layer protocols. This scheme is based on the principles outlined in ISO-TR-9577: protocol identification in the network layer.

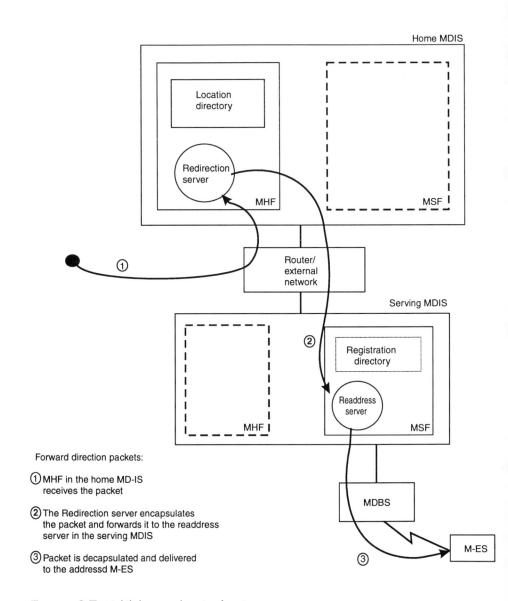

FIGURE 8.7 Mobile home and serving functions.

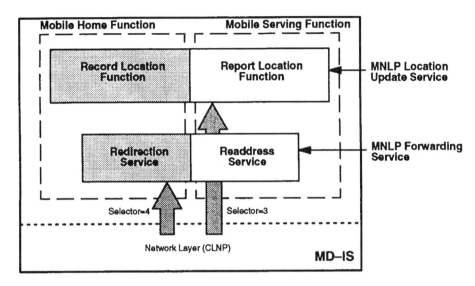

Selector Value	Network Service User
0	CLNP Network Entity
1	TP4
2	MDBS–RRME
3	MNLP Location Update Service
4	MNLP Forwarding Service
5	Channel Status Service

FIGURE 8.8 The NSAP selectors for the MNLP services.

The standards define the first octet of the network service data unit (NSDU) as the subsequent protocol identifier (SPI). The corresponding values for the IP- and CLNP-type NSDUs to be transported via the CLNP protocol are

$$\text{CLNP [2]: SPI} = 0 \times 81$$
$$\text{IP [9]: SPI} = 0 \times CC$$

The corresponding formats of the encapsulated IP and CLNP data packets are shown in Figure 8.9.

The ISO-9577 standards do not enforce a specific scheme of defining the protocol identifier octets. In the case of CDPD, since the inter-MDIS network protocol is clearly defined as CLNP, the equivalent octet, defined as an initial protocol identifier (IPI) is not necessary for the data link service data unit (DLSDU).

Also note that to satisfy the requirement to support IP or CLNP as the network layer to support M-ES to M-ES/F-ES connections, the SNDCP header byte is used.

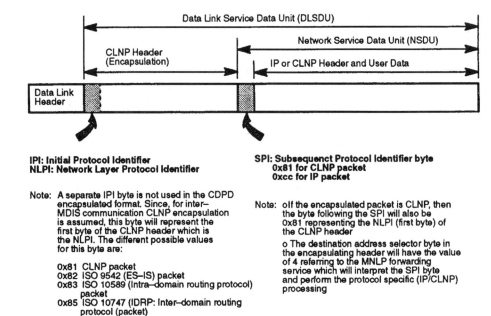

IPI: Initial Protocol Identifier
NLPI: Network Layer Protocol Identifier

Note: A separate IPI byte is not used in the CDPD
encapsulated format. Since, for inter–
MDIS communication CLNP encapsulation
is assumed, this byte will represent the
first byte of the CLNP header which is
the NLPI. The different possible values
for this byte are:

0x81 CLNP packet
0x82 ISO 9542 (ES–IS) packet
0x83 ISO 10589 (Intra–domain routing protocol)
packet
0x85 ISO 10747 (IDRP: Inter–domain routing
protocol (packet)

SPI: Subsequenct Protocol Identifier byte
0x81 for CLNP packet
0xcc for IP packet

Note: o If the encapsulated packet is CLNP, then
the byte following the SPI will also be
0x81 representing the NLPI (first byte) of
the CLNP header
o The destination address selector byte in
the encapsulating header will have the value
of 4 referring to the MNLP forwarding
service which will interpret the SPI byte
and perform the protocol specific (IP/CLNP)
processing

FIGURE 8.9 Encapsulated data formats.

This performs a similar function to the IPI octet specified in the ISO-9577 recommendations.

8.7.4 Information Base Within the MDIS

Three logical groups of tables have been identified as required part of the information base in an MDIS necessary to carry out its functions.

8.7.4.1 *Location Directory*

The information in this directory is required by the MHF for obtaining the forwarding address (the serving MDIS CLNP address) for home M-ESs currently roaming in other areas. Each entry in this table, at minimum, should contain the following:

- M-ES network entity identifier;
- MD-IS forwarding address;
- GMID for multicast M-ES NEIs.

For multicast NEIs, multiple entries in the location directory, one for each unique point of presence will be present. These entries will have unique GMIDs.

8.7.4.2 Registration Directory

This directory contains entries for M-ESs currently registered with this MD-IS. This directory information is used by the MSF function to identify the subnetwork address of the destination M-ES for relay forwarded data packets.

8.7.4.3 Home Domain Directory

The home domain directory (HDD) contains the M-ES and the associated home MDIS addresses. The HDD is maintained by the network administration system of the different network service providers. M-ESs subscribed under a different carrier's network need to be manually added to other carriers' network administration systems where the M-ESs wish to roam. The updated HDD information is expected to be communicated to each MD-IS within that service provider's network. X-500-based HDD entry updates across network administration systems are envisioned for the future. The HDD will typically have multiple entries, each containing the following information:

- Area address;
- Address mask;
- Home area NSAP address of the location service.

The area address and the address mask together determine the range of addresses of the M-ESs (subnet address) whose home MD-IS is defined by the home area NSAP address.

The following sections describe the formats of the MNLP protocol messages and how these messages in conjunction with the MNRP protocol provide the location service and the data forwarding functions that are key to the mobility management within the CDPD network.

8.7.5 MNLP Message Formats

Five different message types are defined as part of the MNLP protocol that are exchanged between the location update NSAPs:

- Redirection Request PDU (RDR);
- Redirection Confirm PDU (RDC);
- Redirect Flush (RDF);
- Redirect Expiry (RDE);
- Redirect Query (RDQ).

The following sections provide brief details of the structure and the encoding of these PDUs that form the data portion of the CLNP NSDUs.

8.7.5.1 Redirect Request Message

The Redirect Request PDU is sent by the serving MDIS to the home MDIS to register the existence and the reachability of the network address of an M-ES. The format of the Redirect Request message is shown in Figure 8.10.

The following parameters can be carried in the Option field.

Group Member Identifier Option

If the source network address represents a multicast NEI, then a 16-bit GMID will have to be specified in this Option field.

Authentication Parameter Option

The source network address needs to be authenticated to screen for validity and provide security from intruders. Presence of this parameter is required for subsequent validation by the home MDIS and granting service access. The format of the Authentication Parameter field is shown in Figure 8.11.

Bit	8	7	6	5	4	3	2	1
Octet 1	PDU Type = RDR (01)							
Octet 2	Registration_Sequence_Count							
Octet 3	Source Address Length							
Octet 4	Source_Network_Address							
•••								
Octet N								
Octet N+1	Forwarding Address Length							
Octet N+2	Forwarding_Network_Address							
•••								
Octet M								
Octet M+1	Options							
•••								
Octet P								

FIGURE 8.10 Format of Redirect Request (RDR) message.

Bit	8	7	6	5	4	3	2	1
Octet 1	Parameter Code = 2							
Octet 2	Length = 11							
Octet 3	Authentication Algorithm 0 = CDPD version 1.0							
Octet 4 Octet 5	Authentication Sequence Number							
Octet 6 •••••• Octet 13	Authentication Random Number							

FIGURE 8.11 Format of authentication parameter option.

Location Information Parameter Option

The service provider affiliation of the currently serving MDIS may be required by the home MDIS for granting service access. The format of this Option field is shown in Figure 8.12.

Bit	8	7	6	5	4	3	2	1
Octet 1	Code = 9							
Octet 2	Length = 6							
Octet 3 Octet 4	Wide Area Identifier (WASI)							
Octet 5 Octet 6	Service Provider Identifier (SPI)							
Octet 7 Octet 8	Service Provider Network Identifier (SPNI)							

FIGURE 8.12 Format of location information parameter option.

Bit	8	7	6	5	4	3	2	1
Octet 1	PDU Type = RDC (02)							
Octet 2	Registration_Sequence_Count							
Octet 3	Destination Address Length							
Octet 4								
•••	Destination_Network_Address							
Octet N								
Octet N+1								
•••	Options							
Octet M								

FIGURE 8.13 Format of the Redirect Confirm message.

8.7.5.2   Redirect Confirm Message

The Redirect Confirm (RDC) PDU is sent by the home MDIS to the serving MDIS as an acknowledgment to the RDR message. The RDC message will contain information that expresses the willingness to provide the data forwarding service or the denial of it by the home MDIS. The format of the Redirect Confirm message is shown in Figure 8.13.

The following parameters can be carried in the Options field:

- *GMID parameter and authentication update parameter.* These two options are similar to those carried in the RDR packet.
- *Result code parameter.* This parameter denotes the acceptance or denial of the forwarding service requested by the RDR message (see Table 8.2).

TABLE 8.2
Result Codes for Redirection Request

Parameter	Meaning
0	Redirection request is accepted
1	No particular reason given
2	MDIS not capable of handling the M-ES at this time
3	M-ES not authorized to use this subnetwork

Parameter	Meaning
4	M-ES gave insufficient authentication credentials
5	M-ES gave unsupported authentication credentials
6	NEI has exceeded usage limitations
7	Service denied on this subnetwork
8–255	Reserved

- *Configuration timer parameter.* This value denotes the value of the configuration timer that the home MDIS prefers the M-ES to use. A zero value indicates that M-ES need not perform a report configuration function and an absence of this parameter indicates that the M-ES should use the default value of 4 hours.

8.7.5.3 Redirect Flush Message

The Redirect Flush (RDF) PDU is sent by the home MDIS to a serving MDIS informing that a network address previously served by this MDIS has moved and that data packets (forward direction) will no longer be forwarded. The format of the Redirect Flush message is shown in Figure 8.14.

The GMID parameter option will be included if the removed network address belonged to a multicast M-ES.

8.7.5.4 Redirect Expiry Message

The Redirect Expiry (RDE) PDU is sent by the serving MDIS to the home MDIS informing that a network address previously served by this MDIS has moved or deregistered and that data packets (forward direction) should no longer be forwarded to there. The format of the Redirect Expiry message is shown in Figure 8.15.

The GMID parameter option will be included if the removed network address belonged to a multicast M-ES.

8.7.5.5 Redirect Query Message

The Redirect Query (RDQ) PDU is sent by the home MDIS to the serving MDIS to query whether an NEI is currently active and reachable via the serving MDIS. The format of the Redirect Query message is shown in Figure 8.16.

The GMID parameter option will be included if the removed network address belonged to a multicast M-ES.

Bit	8	7	6	5	4	3	2	1
Octet 1	PDU Type = RDF (03)							
Octet 2	Registration_Sequence_Count							
Octet 3	Source Address Length							
Octet 4	Source_Network_Address							
•••								
Octet N								
Octet M+1	Options							
•••								
Octet P								

FIGURE 8.14 Format of the Redirect Flush message.

Bit	8	7	6	5	4	3	2	1
Octet 1	PDU Type = RDE (04)							
Octet 2	Registration_Sequence_Count							
Octet 3	Source Address_Length							
Octet 4	Source_Network_Address							
•••								
Octet N								
Octet M+1	Options							
•••								
Octet P								

FIGURE 8.15 Format of the Redirect Expiry message.

Bit	8	7	6	5	4	3	2	1
Octet 1	PDU Type = RDQ (05)							
Octet 2	Registration_Sequence_Count							
Octet 3	Destination Address Length							
Octet 4	Destination_Network_Address							
•••								
Octet N								

FIGURE 8.16 Format of the Redirect Query message.

8.7.6 MNLP Protocol Functions

The location update function and the data forwarding function use the information conveyed by the MNLP message exchanges in conjunction with the MNRP protocol to deliver the services as described in the following sections. The descriptions assume that the home and the serving MDIS of a chosen M-ES are different and that the MNLP messages have to be sent out using the underlying CLNP services. When the home and the serving MDISs are the same, the only difference is that the MNLP messages need not be sent out. Implementation-specific schemes can be used to achieve the same functionality described in the following sections.

8.7.6.1 Location Update

When a serving MDIS receives an End System Hello (ESH) packet from an M-ES, it constructs an RDR PDU with the following information and routes it to the home-MDIS's location update service using the underlying CLNP services:

- The source network address and the registration sequence count are obtained from the ESH packet.
- The home domain directory is referenced to obtain the forwarding network address in the CLNP format.
- Location information parameters (WASI, SPI, SPNI) are obtained from the local database.
- Other optional parameters from the ESH packet, especially the authentication credentials, are copied.

On reception of this RDR packet at the home MDIS, the source network address (the M-ES address) and the registration sequence counts are validated. This is followed by authentication and access control procedures, and a result code is generated. The home MDIS then updates the location directory, restarts the holding timer, and constructs an RDC PDU as a response, containing the following:

- The registration sequence count is obtained from the RDR.
- Updated credential information from the authentication services is obtained.
- The configuration timer value is computed as a fraction (typically half or one-fourth) of the holding timer value for transmission by the serving MDIS to the M-ES.
- The GMID parameter is included if the M-ES NEI is multicast.

The RDC PDU is routed back to the serving MES via the CLNP network layer. Also, if the updated location information denotes a change in the serving area, an RDF packet is constructed and sent to the old MDIS.

On reception of the RDC packet in the serving MDIS, it will construct an MDIS Hello Confirm (ISC) packet from the information extracted from the RDC packet and relay it to the M-ES. The authentication results, updated credentials, and the recommended configuration timer are key fields the target M-ES will be expecting to receive so that data exchanges can be resumed or started.

In addition to the updating location information, the following additional house keeping functions are performed by the location update services.

Flushing Registration/Location Information:

The MDIS on receipt of the Redirect Flush (RDF) PDU, removes the corresponding entry from the registration directory. As seen earlier, an RDF is sent by the home MDIS when it receives the RDR message from an M-ES, which was previously using a different serving MDIS.

Further, when a previously registered M-ES sends an End System Bye (ESB) PDU indicating that it is deregistering from the currently active session, the serving MDIS constructs an Redirect Expiry (RDE) PDU and sends it to the NSAP address of the location update service in the home MDIS of the M-ES identified in the ESB packet. On receipt of the RDE message, the home MDIS removes the location information corresponding to the M-ES identified in the RDE message from the location directory.

Also, if the holding timer corresponding to an entry in the location directory of the home MDIS of an M-ES expires, the associated location entry is flushed.

Query Location Information

This message allows the home MDIS to revalidate the authentication information and the location directory entry of an M-ES. The home MDIS constructs a Redirect Query (RDQ) PDU and sends it to the NSAP address of the location update service in the serving MDIS. On receipt of this message, the serving MDIS examines its registration directory to ascertain the existence of the M-ES identified in the RDQ packet. If the M-ES network address is present, then an End System Query (ESQ) PDU (part of the MNRP protocol) is constructed and transmitted to the subnet address of the M-ES.

This sequence will result in forcing the M-ES to initiate a reregistration by sending an ESH packet that will enable the home MDIS to revalidate the authentication credentials.

8.7.6.2 *Forward NPDU*

When a mobile home function (MHF) receives a data packet for an M- ES not currently located in its home area, it will query its local location directory and identify the NSAP address of the MSF forwarding service associated with the NSAP address of the M-ES. The MHF will encapsulate the data packet as defined in Figure 8.9 and transmit it via the underlying CLNP service.

IF the location directory entry indicates that the M-ES NSAP is a multicast address, the forwarding function will replicate the messages, encapsulate them, and transmit them to the number of unique forwarding addresses found in the location directory.

On error conditions such as absence of a valid entry in the location directory, the data packet (NPDU) is discarded and error-reporting procedures available in the ISO-8473 (ER PDU) can be used to notify the source node of the errors.

8.8 MOBILE NETWORK REGISTRATION PROTOCOL

The mobile network registration protocol is used for exchanging routing and registration information between an M-ES and a serving MD-IS. The protocol itself is connectionless and is designed to operate over a CLNP or Internet protocol (IP).

The MNRP protocol and the MNLP protocol are dependent on each other and differ in the end-to-end entities that terminate the protocol: the MNRP messages are exchanged between an M-ES and its serving MD-IS, whereas the MNLP defines the message exchanges between the serving and home MD-ISs of an M-ES.

8.8.1 MNRP Overview

The MNRP protocol provides configuration information about M-ESs to MD-ISs. Each M-ES informs the MD-IS, the supported NEIs, and the associated subnetwork point of attachment (channel stream identifier and the specific data link connection on that channel stream) using the MNRP protocol.

The MNRP is modeled on the functions specified in the ISO 9542 (end system to intermediate system routing exchange protocol). However, one major difference is that in the case of CDPD, an M-ES can be attached to only one MD-IS—the MD-IS that supports the MDBS/cell/channel stream the M-ES is currently using. Therefore, the redirection subset of the ISO 9542, which assumes that an end system can be attached to multiple intermediate systems—in CDPD there are no better paths to forward NPDUs to their destinations other than the *subnetwork channel* the M-ES is tuned to. Note that the *redirection* functionality is embedded in the MNLP protocol that is executed between MD-ISs.

Reporting configuration of an M-ES (by an M-ES), recording configuration by the MD-IS and the M-ES, flushing old configuration, and querying configuration are the main functions carried out by the MNRP protocol.

8.8.2 MNRP Message Formats

Four different message types are defined as part of the MNRP protocol, which are exchanged between an M-ES and its serving MD-IS:

- End System Hello PDU (ESH);
- End System Bye PDU (ESB);
- MD-IS Hello Confirm PDU (ISC);
- MD-IS End System Query PDU (ESQ).

The following sections provide brief details of the structure and the encoding of these PDUs that form the data portion of the CLNP or IP NSDUs. As we have seen earlier, the M-ES and the MD-IS stacks that support the RF subnetwork can accommodate the CLNP or the IP as the network layer protocol.

8.8.2.1 End System Hello (ESH) Message

The ESH message is sent by an M-ES to register the existence and the reachability of an M-ES (with its NEI) with the MD-IS currently servicing the geographic area in which the M-ES is located. The format of the ESH PDU is shown if Figure 8.17.

The Options field may contain the group member identifier parameter, if the M-ES NEI is of multicast type, and/or the authentication parameters. The authentication parameter format is the same as that described in Section 8.7.5.1 for the RDR message of the MNLP.

8.8.2.2 End System Bye (ESB) Message

The ESB message is sent by an M-ES to the serving MD-IS to deregister the existence and the reachability of its NEI. For a "stationary" M-ES, registration and deregistration strategies will depend on the type of application that is running on the M-ES. M-ESs that stay registered continuously will consume CDPD system resources, but

Bit	8	7	6	5	4	3	2	1
Octet 1	PDU Type = ESH (01)							
Octet 2	Source Address Length							
Octet 3								
•••	Source Network Address							
Octet N								
Octet N+1	Registration Counter							
Octet N+2								
•••	Options							
Octet M								

FIGURE 8.17 Format of ESH message.

will have the advantage of being able to transmit data without any delay; ultimately, the CDPD tariff structure will determine the tradeoff.

The format of the ESB PDU is shown in Figure 8.18.

The ESB PDU may have the GMID parameter in its Option field if the source network address (M-ES address) is a multicast NEI.

Bit	8	7	6	5	4	3	2	1
Octet 1	PDU Type = ESB (02)							
Octet 2	Source Address Length							
Octet 3								
•••	Source Network Address							
Octet N								
Octet N+1								
•••	Options							
Octet M								

FIGURE 8.18 Format of ESB message.

Bit	8	7	6	5	4	3	2	1
Octet 1	PDU Type = ISC (03)							
Octet 2	Destination Address Length							
Octet 3	Destination Network Address							
•••								
Octet N								
Octet N+1	Options							
•••								
Octet M								

FIGURE 8.19 Format of MD-IS ISC message.

8.8.2.3 MD-IS Hello Confirm (ISC) Message

The MD-IS ISC message is sent by an MD-IS to an M- ES to acknowledge the M-ES's existence and to indicate the MD-IS's willingness or denial to provide network service. The format of the ISC PDU is shown in Figure 8.19.

The Options field may contain the Authentication update parameter option, the result code parameter option, and the configuration timer parameter option. These options are the same as those carried by the RDC PDU and described in Section 8.7.5.2.

A zero value in the Result Code Parameter field conveys that the registration was accepted and a forwarding acknowledgment was received from the forwarding function in the home MD-IS of the M- ES. Nonzero values indicate denial of service and the associated codes are defined as part of the RDC message description (Section 8.7.5.2).

The configuration timer parameter of 0 indicates to the M-ES that it does not have report configuration (send ESH PDU) periodically. If this parameter is not present, the M-ES is expected to use the default value of 4 hours as the configuration timer.

8.8.2.4 MD-IS End System Query (ESQ) Message

The MD-IS ESQ message is sent by an MD-IS to validate the existence and the reachability of an NEI at an M-ES. The M-ES is expected to send an ESH PDU on reception of the ESQ PDU. The format of the ESQ PDU is shown if Figure 8.20.

Bit	8	7	6	5	4	3	2	1
Octet 1	PDU Type = ESQ (04)							
Octet 2	Destination Address Length							
Octet 3	Destination Network Address							
•••								
Octet N								

FIGURE 8.20 Format of MD-IS ESQ message.

8.8.3 MNLP Protocol Functions

8.8.3.1 Report Configuration

The M-ES will construct and transmit an End System Hello (ESH) message to inform the serving MD-IS about each active NEI it supports to register the NEI with the CDPD network. The ESH will be sent during the initial establishment of the point-to-point link between the M-ES and MD-IS, when either the configuration timer or the local response timer expires, or when activating an NEI subsequent to the initial data link establishment.

Similarly, when the M-ES decides to inactivate an NEI (i.e., deregister), it will construct and transmit an ESB message. Also, since resources are consumed in an MD-IS during the time the M-ES is known to be active, the M-ES as a matter of procedure should send an ESB message as part of its power-down procedure (before bringing the MDLP link down). If an M-ES fails to deregister, the MD-IS will flush the registration only after the holding timer expires, the default value of which is 8 hours. CDPD implementations may allow the operator to initiate an ESQ message to a chosen M-ES, thereby manually invoking the deregistration procedure.

8.8.3.2 Record Configuration

The record configuration function in the MD-IS, on receipt of the ESH message, extracts the configuration information and stores the {NEI, SNPA} pairs in its local information base. This will replace any previous entry for that M-ES for the same point of presence (defined by the GMID). The location updates in the home and the serving MD-ISs are then performed as part of the MNLP protocol, at the completion of which the serving MD-IS constructs an ISC PDU and transmits it to the M-ES.

On receipt of the ISC PDU, the M-ES stores the authentication update information in the ISC PDU in a nonvolatile storage for use in the next transmission of the

ESH packet. If the extracted result code in the ISC packet is 0, indicating that the registration is successful, the data transfer can commence.

8.9 MOBILE-IP PROTOCOL

The mobile IP protocol is a specification developed by the Internet Engineering Task Force (IETF) Mobile Hosts working group as an approach to transparent mobile interworking [10]. The approach is similar to that of the CDPD system, but the mobile IP is primarily intended for allowing hosts to transparently move from subnet to subnet or from network to network in an established Internet environment. The mobile IP protocol's main objectives are similar to that of the CDPD network:

- The IP address of the mobile system does not change;
- The mobility of a host is transparent to the applications above the network layer and hence no changes will be required for these application layers when mobility is enabled.

The one difference is that the mobile IP does not require changes to routers that do not wish to be involved in the routing of packets to mobile hosts. It requires a special category of routers called mobile support routers (MSR) that have the capability to route packets to "virtual subnets" where mobile hosts reside. In CDPD, the MD-IS that is built as part of the CDPD system conveniently serves as the special router that embodies the CDPD specific routing requirements. CDPD architecture also does not demand existing ISs to have software modifications to support mobility.

A virtual net is composed of many physical segments called cells, which is a network layer concept. Each cell is served by a single MSR. A single MSR may serve multiple cells and multiple cells may reside in the same physical network. A virtual subnet can be thought of as a partitioned physical subnet with the MSRs working cooperatively to heal the partition. The MSRs route packets to and from the mobile hosts (MHs) in their cell(s) and interact with other MSRs to provide the illusion that all the cells are part of the same (virtual) subnet.

As in CDPD, the reverse directional packets do not need any mobility-dependent protocols. Reverse packets from a MH is routed normally through the default router, which is the MSR serving the cell.

For routing packets in the forward direction to a MH, the normal routers may know a route to the associated virtual subnet but will not know which physical cell (and hence the specific MSR) within the virtual subnet is the proper destination for a packet addressed to a particular MH. So a normal router will forward a packet to any MSR it thinks as the best route to the virtual subnet. Two mobile-specific concepts and implementation techniques facilitate the forwarding of the packet to the addressed MH: the mobile interworking control protocol (MICP) and the encapsulation protocol IPIP.

Mobile Interworking Control Protocol

The MICP protocol defines the different message exchanges between MHs and MSRs to accomplish the following functions:

1. *Determination by the MSRs of their serving MHs.* An MSR periodically broadcasts a *beacon* message. The MH that hears the beacon messages from different cells must handshake with one of the MSRs with the GREET MICP message to obtain its service.
2. *Cell switching of MHs.* The serving MSR will continue to assume that the MH is in its cell until it receives a Forward Pointer (FWDPTR) MICP message from another MSR indicating that the MH has moved.
3. *Cold search by an MSR to discover an MH's location.* When an MSR has to forward a packet to an MH whose location it doesn't know, it queries other MSRs using a WHOHAS MICP message. The current MSR serving the MH will reply with a IHAVE MICP message.

Encapsulation Protocol IPIP

An MSR that receives a packet for an MH will determine by the MICP protocol facilities the nonvirtual-subnet address of the MSR that is currently serving the MH and forward the packet to that MSR, using the reset of the Internet as a delivery mechanism. These packets are tunneled between MSRs by encapsulating IP within IP.

The above description strictly applies to intercampus routing. If an MH moves to another campus, the MH is made to handle its own tunneling to and from one of the MSRs in the home campus and is called *popup MH*. These MHs temporarily acquire an address, called the *nonce address*, which should be in the virtual subnet of the foreign campus to facilitate the popup's movement in the foreign campus. Also, the popup communicates the nonce address to an MSR in its home campus via the MICP popup greeting message; this MSR will behave as if the popup were one of its local MHs. Therefore, the popup has a home address by which it is known to the world and has a nonce address that it uses to access local routing services in a foreign campus.

In comparison to CDPD, if we assume that the campus setup in a mobile IP framework and an area (served by an MD-IS) in the CDPD framework are roughly equivalent, we can observe the following:

- The home and serving MD-IS concepts in CDPD are roughly duplicated in the mobile IP for cases where a roaming M-ES is communicating with an F-ES serviced by an MD-IS different from that currently serving the M-ES. That is, in the mobile IP case, the forward direction packets have to first go to an MSR in the home campus before being forwarded to the MH.

- The mobile IP, however, has a more efficient scheme for the forward direction packet routing if the source that is sending the packet to the MH is also in the same campus. In this case, the local routing entities are aware of the MH's temporary address and the packet can be forwarded locally, and the packet does not have to take its route via the home campus. Whereas, in CDPD, the forward packets have to be routed via the home MD-IS even if the M-ES and its communicating peer are both located in the currently serving MD-IS area. This can be seen as a major drawback in the CDPD routing structure. However, the advantage is that the routing protocols (MNRP/MNLP) are simpler in CDPD and incur relatively low overhead.
- Intra-area routing in the CDPD is extremely simple compared to the intracampus routing. The mobile IP uses complex MICP and still may have to use data encapsulation. In CDPD, from the MD-IS's point of view, the M-ESs are technically connected in a *star* configuration and internal routing with the SNPA (<cell, channel stream>) routes the packet to the associated M-ES.

We can observe that the wireless aspect of the CDPD system, the architectural differences in the networks for which the mobility protocols are being used, and the mobility characteristics of the devices intended to be used in these different networks may account for the differences in the protocol structures. Also, since the MD-IS is delivered as part of the CDPD system deployment, all mobile specific protocols are localized to CDPD components and no changes to the existing routers are necessary.

References

[1] CDPD Specifications Release 1.1, CDPD Forum.
[2] ISO-8473: Information Processing Systems—Data Communications Protocol for Providing the Connectionless-Mode Network Service and Provision of Underlying Service.
[3] Perlman, R., *Interconnections: Bridges and Routers*, Reading, MA: Addison Wesley Publishing Company, Professional Computing Series, 1992.
[4 Piscitello, D. M., and A. Lyman Chapin, *Open Systems Networking: TCP/IP and OSI*, Reading, MA: Addison Wesley Publishing Company, Professional Computing Series, 1993.
[5] ISO-10747: Information Processing Systems—Telecommunications and Information Exchange Between Systems—Protocol for Exchange of Inter-Domain Routing Information Among Intermediate Systems to Support Forwarding of ISO 8473 PDUs.
[6] RFC-1467: Topolcic, C., Status of CIDR Deployment in the Internet, August 1993.
[7] ISO 10589: Intermediate System to Intermediate System Intra-Domain Routing Information Exchange Protocol for Use in Conjunction With the Protocol for Providing Connectionless-Mode Network Service (ISO 8473), International Organization for Standardization (ISO), 1992.
[8] ISO 9542: End System to Intermediate System Routing Exchange Protocol for Use in Conjunction With the Protocol for Providing the Connectionless-Mode Network Service (ISO 8473), International Organization for Standardization (ISO), 1991.
[9] RFC-791: Postel, J. B., Internet Protocol, 1981.
[10] Ioannidis, J., D. Duchamp, G. Q. Maguire, and S. Deering, "Protocols for Mobile Internetworking," Mobile Hosts Working Group, Internet Draft Proposal.

CHAPTER 9
▼▼▼

NETWORK MANAGEMENT

9.1 INTRODUCTION

The successful operation of a public mobile data communications network like the CDPD network depends, to a large extent, on the availability of additional services that assist the CDPD service provider in managing the network. These services perform functions that do not directly impact the data carriage services for the subscribers of the CDPD network. The services that would fall into this category would be those of configuration, operations, accounting, and a few other services that go into building a commercially deployed network. These services are generally bundled together as a package of network management services. In addition to network management services, these functions allow collection of network usage information to support billing and protect the service provider and subscribers from unauthorized usage of the network resources.

9.2 OVERVIEW OF CDPD NETWORK MANAGEMENT

The network management services support service providers by providing the capability for fault, configuration, performance, and security management functions. These functions are essential to the commercial operation of the CDPD network services.

The CDPD specification was intended as an open standard, and the service providers may purchase components from various suppliers and manage them using a common management framework. Different service providers can exchange CDPD management information by following the common management specifications.

A CDPD service provider's network is managed using X.700 and, optionally, SNMPv2. The X.700 management protocols and functions are specified by the Network Management Forum set of OMNIPoint 1 documents. These include the common management information protocol (CMIP) and the common management information service (CMIS). In addition to the X.700, M-ESs in the CDPD network are managed using SNMPv2.

9.3 CDPD MANAGEMENT FUNCTIONS AND CAPABILITIES

A CDPD network management system (NMS) will offer network management services in the functional areas outlined below. Note that services in all of these areas are necessary to completely manage a service provider network.

- Configuration management;
- Fault management;
- Performance management;
- Security management;
- Accounting management.

9.3.1 Configuration Management

The CDPD NMS configures the various physical and logical elements of the service provider network. These elements will include the MDBSs, the MD-ISs, the mobile data link connection, the cell, the channel streams, the RF channel, and other network elements (like the IS) that go into building the physical network. The configuration of the accounting server is handled in the section on accounting management.

Configuration in the context of this network implies exercising control over, identifying, collecting data from, and providing data to network elements.

A real-world network will go through the following steps before it is operational.

- *Installation.* This is where the physical subsystems are located and assembled together, and the necessary power and other wiring needed is connected up. Typically, no configuration is required at this stage.
- *Provisioning.* This consists of procedures necessary to bring each piece of equipment into service. Once the unit is ready for service, the supporting pro-

grams are initialized. The state of the unit (e.g., in service, maintenance, out of service) can also be controlled by the use of provisioning functions. Each of the network elements is configured with the required parameters, new elements may be defined, connection to other network elements may be defined, and the equipment can be restarted. The network element's database is usually initialized at this stage. Addition of cells, channel streams, and RF channel configuration is done at this step.

- *Operational Status and Control.* The network can be commissioned into operation after current status, statistics, network alarms can be monitored and current configuration and parameters can be modified. The network element status and control includes the following functions (this list is not exhaustive and is representative of the kinds of functions required by operations personnel in a commercial data communications environment):

  - Report current status;
  - Schedule status reports;
  - Set service availability timetable;
  - Allow/inhibit automatic restoration;
  - Initiate/release change to backup;
  - Report automatic restoration status;
  - Report status of message store and forward facilities;
  - Report status of circuits and transmission equipment;
  - Acceptance testing including software staging and invocation.

9.3.2 Fault Management

This category of network management functions includes the monitoring of exception conditions such as equipment or software failures, connectivity breaks, and other functions intended to maintain operations in the face of unusual situations. They also include routine testing functions intended for system maintenance. In addition, the broad category of fault management encompasses fault correction and recovery procedures.

Fault and failure conditions are required to be detected for all CDPD components (both logical and physical components) that directly contribute to providing the CDPD network services. The detection of faults leads to the generation of alarms or notifications. Alarms are also generated on the correction of the fault.

9.3.2.1 Alarm Management

Alarms are issued by the network elements under the following conditions:

- An equipment failure occurs or is cleared.

- A software or processing failure occurs or is cleared. The following examples illustrate this condition:

  1. An accounting server fails to receive traffic matrices from any of its client MD-ISs.
  2. A processor runs out of its available memory resources.

- A communications failure occurs or is cleared.
- There is a change in the quality of service provided. As an example consider the following cases:

  1. Excessive block error rate (BLER) detected on a channel stream.
  2. Highest congestion level on a network element for a configured duration of time.

- The operational state of the network element changes. The following examples illustrate this scenario:

  1. A network port may go to an "out-of-service" state.
  2. A processing element fails to report its status and is thus presumed to be in the "failed" state.
  3. A RF channel is put into "maintenance" state for repair purposes.

- Contact with other network elements has been lost. The following examples illustrate this scenario:

  1. Contact with a neighbor MD-IS is lost.
  2. Contact with a cell site MDBS has been lost.

- A security threat is detected. The following examples demonstrate this condition:

  1. The authentication server detects repeated failures in authenticating a particular NEI (in most deployed CDPD networks today, the NEI is identical to the IP address, consisting of 4 octets).
  2. Repeated failed attempts of a possible operator to authorize access into the network management system with a password.

9.3.2.2 Fault Correction

Most of the CDPD service providers, based on their telecommunications and cellular wireless experience, have required network infrastructure vendors to implement substantial redundancy. Typical service affecting critical failure situations are handled automatically by switching in a redundant component without any manual management intervention. Some failure conditions may require operator intervention. However, configuration of the redundant components in order to designate the backup components, switching back to the primary components, or a complete reload of the system, is normally controlled by explicit operator management action.

Most of these fault correction functions apply to the following service-critical areas:

- Physical components:

  1. Automatic switching in of a backup CPU or processing element.
  2. Disk mirroring.
  3. Redistribution of load among multiple available processors on the failure of one of them.
  4. Switching in of a backup power supply.
  5. Switching in of a hot standby clocking unit.
  6. Automatic switching in of a network access unit (e.g., Ethernet interface card, T1 interface module).

- Functional elements:

  1. Automatic switching in of a hot standby RF sniffer element.
  2. Alternate RF channel switching in.
  3. Automatic backup accounting server switch-in.
  4. Manual switch in of a backup authentication server.

- Communication circuits:

  1. Backup backbone router path redundancy.
  2. Backup T1 circuit.
  3. Backup operator console access to the network management system.

9.3.3 Performance Management

This aspect of network management deals with collection of performance metrics and invoking performance controls.

Performance data collection is required to ensure that the CDPD service provider network is delivering the expected quality of service. This data, after analysis, provides useful input on both the short-term performance processing and the long-term trends. Some of the configuration parameters can be modified based on the analysis of the collected metrics. This in turn contributes to overall network planning. In addition, some parts of the collected data may even be used for fault or security management.

The types of analysis that may be performed on performance data will include the following:

- Identification of congested elements (e.g., congested channel streams, cells, MD-IS to MDBS connections, authentication server, accounting server).
- Identification of the busy hour, at the cell level, for CDPD traffic. This can help in planning additional traffic channels or increasing bandwidth on the MD-IS to MDBS circuit.

- Identification of cells with suboptimal throughput levels. This can help resolve potential interference issues with the analog system or other RF conflict.
- Identification of processors within a network element that are performing outside of expected bounds (higher or lower).
- Identification of RF channels with low or high availability. If an RF channel is not available most of the time for CDPD usage, and if the channel utilization for CDPD is high, the number of AMPS channels allocated to CDPD are not adequate for this cell.
- If collection of metrics on an individual M-ES basis, at the MDBS, is turned on, it can provide useful input in case of problems noticed by the M-ES user. At the MD-IS, if performance metrics are collected on a per data link connection level, counters like "number of reassembly errors," "number of unsuccessful registrations," and so forth can lead to useful conclusions, based on which the service provider can tune network parameters.

9.3.4 Resource Management Functions on CDPD Network Elements

The model laid out for resource management within the CDPD network context deals with the management of functions provided by the MD-IS and the MDBS components of a CDPD service provider network without regard to the physical hardware elements that implement the MD-IS and the MDBS functions. The network management functions have been abstracted to a level that requires knowledge about the CDPD network and services, but does not necessarily draw upon the knowledge of the equipment that goes into building the network.

Presented below are some of the more significant management functions that can be performed on the resources of the MD-IS, the MDBS, MDBS connections, channel stream, mobile data link connection, and RF channel.

9.3.4.1 *MDBS Resource Management*

This will include all the functions necessary to configure an MDBS and the various resources that an MDBS controls, including the following:

- Install/deinstall MDBS.
- Change MDBS configuration parameters:
  1. Change operational parameters (e.g., MDBS NEI);
  2. Change event reporting or logging criteria;
  3. Change performance data collection elements or interval;
  4. Control whether or not the MDBS may carry traffic;
  5. Configure distribution intervals for the various RRM broadcast messages.
- Configure cell:
  1. Requires adjacent cells to be redefined;

2. May require the cell to be disabled for carrying traffic, in case of removal of the cell.

- Install channel stream.
- Control channel stream—allow/disallow traffic, forcing a channel hop.
- Configure RF channel.
- Manually direct a channel hop.
- MDBS fault management:

    1. Detection of MDBS faults;
    2. Localization of the fault;
    3. Fault determination.

- Run MDBS self-test.
- Report RF channel measurements.
- MDBS performance management:

    1. Control MDBS performance reporting;
    2. Report MDBS performance metrics.

9.3.4.2 MD-IS Resource Management

This includes all the functions that provide for the configuration of an MD-IS and the various resources that an MD-IS controls, including the following:

- Install or deinstall MD-IS.
- Configure MD-IS.
- Install or deinstall MDBS connections.
- Control mobile data link connection:

    1. Reporting its attributes;
    2. Exert flow control mode to the relevant M-ES.

- Report currently registered subscribers.
- MD-IS fault management.
- Report MD-IS alarms.
- Run MD-IS self-test.
- Perform M-ES loopback test:

    1. Done using the MDLP specified TEST function;
    2. Report the results of this test.

- MD-IS performance management:

    1. Set MD-IS performance reporting interval;
    2. Report MD-IS performance metrics.

9.4 SECURITY MANAGEMENT

Security management of a network consists of two areas: security of management and management of security.

9.4.1 Security of Management

The primary mechanism for ensuring the security of the CDPD management network is through the use of secure communications. This can normally be achieved through a service-provider-controlled data network. Some form of authentication information may be exchanged at the time of initial association establishment (between the different connection management entities—the CMEs) and during the lifetime of an association. Specific requirements for security of management are normally going to be determined by each service provider.

Password-protected access into the network management system is usually present on the managing system. There may be multiple levels of password protection, depending on the specific action to be performed on particular resources. As an example, an action requiring an operator to add a new cell site into the system may require the operator to enter the highest level of password since this action is potentially intrusive to the network operation and requires a complete understanding of the current state of the network and the actions for backup and recovery on failure, if any. On the other hand, a report generation of "the last 24-hour network statistics" may only require the operator to enter a "low" level of password, though different service providers may view even this action as a security breach and may require only system administrator access for this function. The network management system (NMS) may monitor invalid access into the system and report these as security alarms.

9.4.2 Management of Security

This is a general problem not unique to CDPD. Service providers are expected to analyze their own security requirements and points of vulnerability. The mechanisms used to secure the airlink, namely, the authentication and encryption across the airlink, are dependent to maintaining the confidentiality of the secret keys. It is worth mentioning here that the mechanisms providing confidentiality over the airlink only protect the subscriber and the subscriber's data from casual eavesdropping. As is the case with any commonly used security mechanism, the knowledge of the security system and development of specialized tools can aid the skilled hacker to break into this network.

Based on long and painful experience with security holes in the existing cellular systems, the CDPD industry has worked towards special measures to combat security attacks. These measures are in addition to the preventive mechanisms that the CDPD specification has laid out. These measures can be classified into the following categories:

- Network-side procedures;
- M-ES-implemented measures.

9.4.2.1 *Network-Side Procedures*

The network management system can track security attacks on the network. This tracking can be based on logic like "generate an alarm after some number of unsuccessful registration attempts with wrong credentials sent from an M-ES" or even "generate a report of all the detected duplicate NEIs." Under network operator or automatic control, inquiry messages (the ESQ message of the MNRP protocol) may be sent to all or some subset of the M-ESs to identify and authenticate them. A message may be sent to an unauthorized M-ES to disable (Zap) an M-ES. Some service providers have required the MD-IS to generate a new encryption key at a fairly high frequency, like once every half-hour, in order to reduce the probability that a *clone* M-ES will be able to use the network for very long on an earlier encryption key that the clone had become aware of. The configuration timer has also been adjusted to force M-ESs to reregister at regular intervals, on the order of once every hour. The MD-IS may, however, assign a new set of credentials only every so often (i.e., the hacker M-ES may not even be aware of the fact that a new set of credentials is to be generated with a specific registration attempt). The reverse data traffic is not passed onto the backbone router network at the MD-IS unless the source M-ES NEI is currently registered and has therefore passed the authentication test. The network can support restrictions on access by or to different NEIs, such as restrictions by location or geographic areas.

All the measures above only reduce the probability of attack on the network's security system and mitigate the amount of damage to the subscribers of this network. Of course, with the advance in knowledge, there will be potential *CDPD scanners* out there that can capture and understand each and every byte of information going across the airlink and thereby work around any of these measures.

These procedures are in addition to the security measures usually adopted for the protection of a service provider's data communications network. These default measures can include router firewalls, trusted access into the network management system, multiple levels of password protection, allowing a particular M-ES to only talk to a limited number of configured end systems (identified by NEIs that are part of a screening list), disabling telnet, finger and rlogin access into the MD-IS.

9.4.2.2 *M-ES–Implemented Measures*

An M-ES may require the user to enter a personal identification number (PIN) or other form of password before allowing any usage of the device. The M-ES network management agent (e.g., SNMP), if implemented in the device, may also be able to generate a security alarm after some number of invalid attempts (which may be configurable by the service provider) and then disable itself. Some networks may allow the M-ES to use different encryption and authentication algorithms, this being provided for by the CDPD specification. A point to be noted in this context is the fact that the CDPD specification does not provide for any services that will support bilateral authentication across the airlink. This means that the CDPD network is not validated to the M-ES across the airlink.

Applications making use of the M-ES can implement end-to-end encryption if the demands of the application justify the added expense of data confidentiality over the entire data communication pipeline.

9.4.3 Airlink Security

The primary set of security functions that are supported over the CDPD Airlink Interface are as follows:

- *Confidentiality of the data link*. After the secret keys have been determined, all information going across the airlink, in both directions, is encrypted. This function of encryption is performed by the SNDCP protocol layer at the M-ES and its serving MD-IS as part of the construction of the SN-DATA PDUs.
- *Key management*. The encryption algorithms that support the confidentiality of the information across the airlink work with secret keys. The secret keys are exchanged between the M-ESs and the serving MD-IS. The management of the secret keys is done by the network.
- *M-ES authentication*. Each of the NEIs used by the M-ES is authenticated by the CDPD network to ensure that only the authorized NEIs are allowed usage of the airlink. The exchange of authentication data takes place between the M-ES and its serving MD-IS. The serving MD-IS and the home MD-IS exchange authentication data according to the mobile network location protocol (MNLP).

9.4.3.1 Key Exchange

Key exchange procedures are performed by a security management entity (SME) resident in each M-ES and in the MD-IS. The SME functionality in the M-ES is supported in the subscriber identity module (SIM) if the SIM is implemented as a detachable module. The SME is a layer 3 (network layer) entity that is accessed using the SNDCP layer. The SNDCP layer provides the primitives of SN-UNITDATA with the acknowledged class of service to the SMEs for the purpose of key exchange.

Key Management

Key management is based on the electronic key exchange procedure of Diffie and Hellman. This procedure is executed after initial establishment of the data link connection but before the NEI registration and authentication. This procedure is re-executed from time to time to change the encryption keys used by the data link encryption algorithm. Key resynchronization takes place after every re-establishment of the data link connection.

Key Exchange Algorithm

The base algorithm requires the M-ES to generate a secret random quantity x and the MD-IS to generate another secret random quantity y. The actual generation of these random values is implementation-dependent. The size of the random quantities is also implementation-dependent, but does not exceed 256 bits, nor fewer than 40 bits. Smaller sizes make the algorithm execute more rapidly, but correspondingly reduce the security of the procedure since the probability of successfully deriving the keys by computation goes up. Typically, the network management system will allow the option to the service provider to configure key sizes. CDPD deployment in other countries will most likely have key sizes restricted to 40 bits, since software containing a key generation algorithm of more than 40 bits is not allowed to be exported from the United States to other countries.

The MD-IS also generates two public quantities, a base a, and a modulus p. The value p is a prime number larger than a. Both a and p are 256 bits long. Security-conscious networks have required the MD-IS to be able to change the values of a and p from time to time.

The serving MD-IS initiates and controls the execution of the key exchange algorithm. The initiation of the algorithm is by transmitting the IKE (the MD-IS key exchange PDU), consisting of the triplet of $(a, p\ a^y \bmod p)$. The M-ES replies by transmitting the EKE (M-ES key exchange PDU) consisting of the quantity $(a^x \bmod p)$ to the MD-IS. The M-ES and the MD-IS will now both generate $(a^{xy} \bmod p)$, which is a shared secret quantity. All of these numbers are 256 bits long. This sequence of the key exchange handshake between the M-ES and the MD-IS is illustrated in Figure 9.1.

Secret Key Derivation

The M-ES and the serving MD-IS then derive a pair of shared secret keys (k_0, k_1) from the quantity $(a^{xy} \bmod p)$, using procedures specific to the data link encryption algorithm of choice. The supported algorithms are RC4 and the TEST. Support for other encryption algorithms (e.g., DES, CMEA) is available in the IKE PDU format.

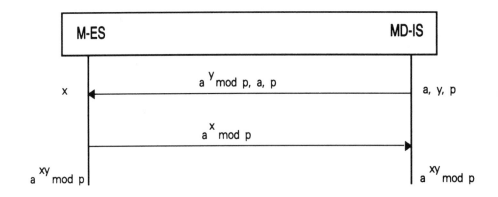

FIGURE 9.1 Key exchange sequence.

The secret key k_0 is used to encrypt data at the MD-IS and decrypt received data at the M-ES (this key is used for the forward direction transmissions). The secret key k_1 is used to encrypt data at the M-ES and decrypt received data at the MD-IS (this key is used for the reverse direction transmissions). Thus, k_0 is the encryption key for the MD-IS and the decryption key for the M-ES, while k_1 is the encryption key for the M-ES and the decryption key for the MD-IS.

9.4.3.2 Data Encryption and Decryption

After the key exchange procedure has been completed, all point-to-point information transfer over the acknowledged data link service is encrypted. This includes the NEI registration procedure after the initial key exchange procedure. Broadcast and multicast information transfer is not encrypted.

The pair of shared keys is used to encrypt the DATA SEGMENT fields of all SN-DATA PDUs transmitted by the M-ES and the MD-IS. The encryption algorithm currently supported by all CDPD networks is the RC4 algorithm.

RC4

The RC4 algorithm is a stream cipher that generates a stream of pseudorandom data from the key. This stream is known as the *keystream*. The algorithm is initialized on each change of the key and is reinitialized whenever the data link is re-established. Separate keystreams are generated for encrypted data by the MD-IS on the forward channel and by the M-ES on the reverse channel.

Each consecutive bit of the keystream is exclusive-ORed with a bit of data to be encrypted in order of bit transmission. After the last bit of the Data Segment field of an SN-DATA PDU is encrypted, the next bit of the keystream is applied to the first bit of the next SN-DATA PDU. As a convenience of implementation, this process is executed one octet at a time.

Data is decrypted by applying the same process to the received SN-DATA PDUs. The RC4 algorithm generates the same keystream that was used to encrypt data at the transmitter. Each bit of the keystream is exclusive-ORed against the received corresponding encrypted data bit.

TEST Mode

This algorithm is reserved for test and experimental purposes. During the initial deployment of the CDPD networks, this mode was used to indicate that encryption would not be used for the information transfer. This was useful to isolate other problems with the network since it allowed the systems integrators to capture and analyze data sent over the airlink.

9.4.3.3 *Authentication*

Authentication procedures are needed to validate the NEI registration. These procedures are performed using the mobile network registration protocol (MNRP). The MNRP procedures are typically implemented as a management entity, resident in each M-ES and in the MD-IS.

The concept used in the M-ES authentication is based on the notion of establishing a shared historical record of all interactions between the M-ES and the network. There is no permanent secret, like a password, that the M-ES possesses. This notion provides protection against clones that might be created by theft of a permanent secret.

The authentication procedure is performed as part of the NEI registration process, after initial establishment of the data link connection and the establishment of the shared secret key for operating the encryption algorithm. The data that goes into the authentication handshake is encrypted according to the shared secret keys. The MD-IS, based on the configuration set up by the service provider, may require the M-ES to repeat the authentication procedure from time to time.

9.5 ACCOUNTING MANAGEMENT

9.5.1 Overview of CDPD Accounting

The cellular digital packet data accounting mechanism provides information about how the CDPD network resources are being used. The key resource that is being accounted for is the airlink resource. Since the CDPD network provides connectionless datagram services to M-ESs, the accounting approach is to accumulate statistics about network layer protocol data units (NPDUs) that cross the mobile data link.

The CDPD accounting model comprises an accounting meter, which resides in the MD-IS, and an accounting server, which resides in an external machine independent of the MD-IS. The network management entity defines when and how frequently the accounting meter is expected to report the traffic matrix segments (collection of the statistical information on the resource usage) to the accounting server.

The CDPD accounting mechanisms in themselves do not address pricing, billing, the reconciliation of usage claims among CDPD network service providers, or receivable processing. These functions are accomplished by various accounting applications. The main purpose of the CDPD accounting is to capture and distribute the usage of the CDPD resources so that accounting applications (e.g., billing applications) have the necessary data to perform their functions.

9.5.2 Accounting Model

Accounting meters in the serving MD-ISs capture the NPDU statistics and deliver these statistics (using any of the accounting protocols—X.400, ftp, rcp, etc.) in the form of a traffic matrix segments (TMS) to a serving accounting distributor (SAD). The SADs in turn sort the traffic matrix segment rows (received from all the accounting meters within the service providers network) according to the location of the accounting home (defined in the subscriber profile database) for a particular mobile subscriber, and deliver home accounting segments (HAS) to the appropriate home accounting distributors (HAD). Serving accounting distributors also deliver a copy of the home accounting segment to a serving accounting collector (SAC) (the SACs serve as a repository for the service providers). The HADs receive the HAS flows from different SADs, sorts the HAS flows, packages them into a consolidation accounting segments (CAS), and delivers these segments to the home accounting collector (HAC) and to the consolidation accounting collector.

The CASs collected at the CAC are then sent to the billing application, which may or may not reside in the same system. In a typical accounting model, the CASs are usually delivered to another system to perform the billing functions.

The following sections detail the accounting meter and the accounting server of the CDPD accounting model at a high level.

9.5.3 Accounting Meter

An accounting meter (AM) is an entity that collects statistical information on the usage of the CDPD resources in the service providers network by the MESs. Conceptually, there is an AM in each MD-IS. It collects information on M-ES registration, deregistration, and NPDU traffic and periodically reports them as a traffic matrix segment to the SAD.

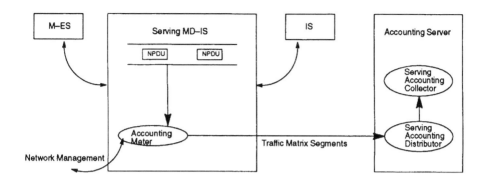

FIGURE 9.2 The accounting model.

The datagrams, NPDUs flow though the MD-IS between MDBSs and the intermediate systems (ISs). These NPDUS, registration messages, and deregistration messages are counted by the AM in the MD-IS and packaged into traffic matrix segments and are reported to the SAD (discussed in the preceding sections of this chapter) periodically. At the end of each reporting interval (a configurable parameter), the AM captures the content of the traffic matrix in a traffic matrix segment. Simultaneously with this action, a new traffic matrix is initialized by the AM and statistical information continues to be collected on registration/deregistration events and the flow of NPDUs for the following traffic matrix segments.

Figure 9.2 illustrates the accounting meter and its interface with the accounting server at a high level.

The AM delivers the resource usage information in the form of a segment called the traffic matrix segment (TMS). A TMS contains a header and many traffic matrix row entries of identical structure (but different information). Figure 9.3 shows the format of a TMS.

Each TMS row entry can be considered as a record. Each record contains information of a session between two network entities (e.g., between an M-ES and F-ES or between two M-ESs) over a particular period of time (AM duration). The header captures the information that is common to all the row entries, including the start time and actual duration.

The row entry captures the information about the NEIs, the service providers tariff code, data type, and so forth.

9.5.4 Accounting Server

An accounting server provides a secure, reliable repository for information collected from all AMs and also distributes the collected information to the appropriate billing system to perform the final billing for the service provider's subscribers. A typical

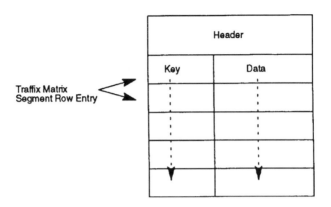

FIGURE 9.3 Format of a traffic matrix segment.

accounting server comprises of a SAD, a HAD, a SAC, a HAC, and a CAC. This chapter deals with all the subsystems involved in the accounting server and how they interface with each other and also with the billing system.

Figure 9.4 illustrates the accounting server and its subsystems.

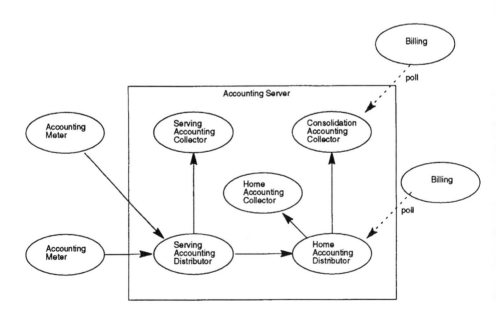

FIGURE 9.4 Accounting server.

9.5.4.1 Serving Accounting Distributor

The SAD provides near-real-time accounting functions in order to track the resource usage by a particular subscriber. Each subscriber when provisioned would contain the home accounting location, which may or may not be associated with that of the home MD-IS. The home accounting location in the accounting server refers to a home accounting distributor (will be discussed later). Each HAD would be assigned an X.400 O/R address, which is a part of the subscribers database. Hence, the SAD uses the subscriber's NEI to determine the appropriate HAD.

The SAD accepts traffic matrix segments sent to it by different AMs within the service providers network. The individual traffic matrix segments are then collected into a serving accounting flow by the SAD and processed.

The SAD has the option of modifying/replacing the serving tariff code in each row entry of a traffic matrix segment. This allows the serving service provider network to include tariff information in the accounting row entries prior to distribution to the home accounting locations. The serving accounting distributor is illustrated in Figure 9.5.

Periodically, the SAD sorts the serving accounting flow into home accounting segments. The traffic matrix segments are sorted according to the MES NEIs (a part of the traffic matrix segment data) destined to particular HAD and consolidated into segments called home accounting segments. The MES NEI is given as part of the traffic matrix, which is used as the key search parameter into the local table to determine the HAD (i.e., O/R address of the HAD).

Each row in a traffic matrix segment, shall be contained in exactly one (no more and no less) home accounting segment and distributed to a HAD and a SAC.

Home Accounting Segment

A typical home accounting segment consists of a home accounting header and potentially multiple home accounting segment rows. Figure 9.6 illustrates a typical home accounting segment.

9.5.4.2 Serving Accounting Collector

The SAC provides functions supporting the storage, analysis, and processing of the accounting information collected within its own serving area. The exact nature of the functions provided on the SAC is specific to the service provider's needs. A typical SAC would deposit the home accounting segments into a database, which will then be processed by a service provider's application. Usually, the SAC is good place to locate if the home accounting segments are being received from the SAD and if the information in the home accounting segments are accurate if there seems to be a problem in the flow of accounting information (since all the subsystems of the accounting server do not have to reside in the same machine).

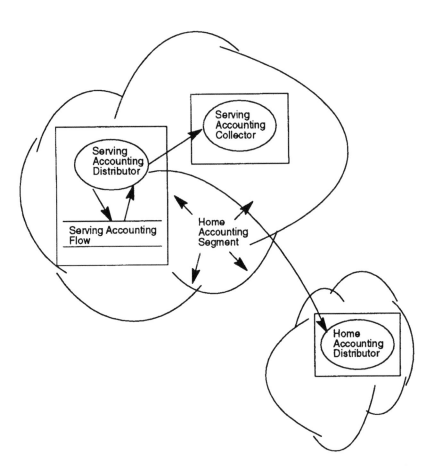

FIGURE 9.5 Serving accounting distributor.

9.5.4.3 Home Accounting Distributor

A HAD is the last leg of the accounting server before the data is packaged and transmitted to the billing system. A HAD accepts home accounting segments from many
(potentially) multiple SADs and distributes them to the CACs and to the HACs. The
HAD finds the O/R address of the appropriate CACs or HACs from the subscriber
profile maintained by directory services. Figure 9.7 illustrates a home accounting
distributor.

A consolidation accounting segment contains all of the information that was in
the home accounting flow. This segment is sent to the HAC. The HAD also sends to
the HAC a copy of each of the consolidation accounting segments, which it distributes to the CACs.

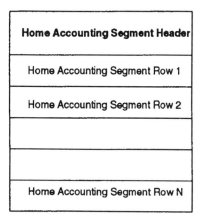

| Home Accounting Segment Header |
| Home Accounting Segment Row 1 |
| Home Accounting Segment Row 2 |
| |
| |
| Home Accounting Segment Row N |

FIGURE 9.6 Home accounting segment format.

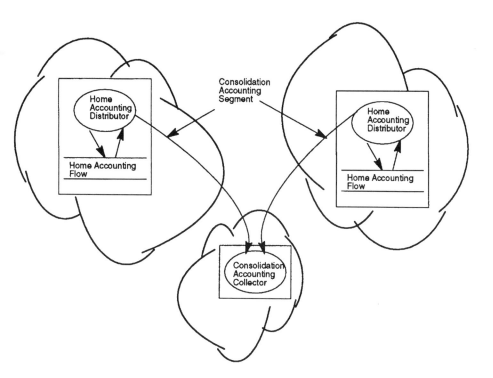

FIGURE 9.7 Home accounting distributor.

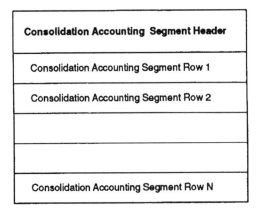

FIGURE 9.8 Consolidation accounting segment.

Consolidation Accounting Segment

A typical consolidation accounting segment consists of a consolidation accounting header and potentially multiple consolidation accounting segment rows. Figure 9.8 illustrates a typical consolidation accounting segment. The address of the consolidation accounting collector for a mobile subscriber is part of that subscriber's profile, which is maintained in the directory service. The CAC is represented as an X.400 O/R address.

9.5.4.4 Home Accounting Collector

A HAC collects consolidation accounting segments from its HAD. This includes a consolidation accounting segment that contains all information in all home accounting segments collected during the home accounting distributor span. Segments received by a HAC also include a copy of each consolidation accounting segment that was sent to the CACs. The HAC provides certain functions to process the consolidated accounting segments. The nature of the functions supported at the HAC depends on the service provider and their requirements. A typical service provider would analyze these consolidated accounting segments for billing purposes and also for interservice provider-network billing reconciliation.

9.5.4.5 Consolidation Accounting Collector

A CAC is a special kind of HAC that collects consolidation accounting segments from multiple HADs from multiple service provider networks. The figures in Section 9.5.4.3 illustrate the CAC's relationship with the HAD and other entities.

The CAC stores the consolidated accounting segments into a database or delivers these segments to a billing provider in order to prepare the bill for the subscribers. These consolidation accounting segments collected at the CAC are also used for interservice provider-network billing reconciliation.

9.5.5 Accounting Meter Management Attributes

In this section, we will discuss one of the main AM objects and its management. One of the main AM objects is the timestamp in the traffic matrix. Accounting meters need an accurate source of time in order to generate the meter start time and meter sent time. Also, the AM shall record the current date and time of each registration/deregistration event in the traffic matrix.

Each MD-IS on behalf of its AM is required to maintain accurate time of day. All times used in CDPD are to be reported in Greenwich mean time (GMT), also known as coordinated universal time (UTC), with a precision of at least 5 seconds. (That is, the time maintained by an accounting server may differ from true GMT by no more than plus or minus 5 seconds).

The mechanism of maintaining time-of-day clock synchronization with a recognized authority is not defined by CDPD, but is left as an implementation decision. There are a number of methods available for the service provider to choose from:

- A WWV radio receiver;
- A dialup line 1,200-bps modem and software to synchronize with the time services operated by the National Institute of Standards and Technology or the U.S. Naval Observatory;
- Software to use a time server and a time protocol such as simple network time protocol (SNTP) or network item protocol (NTP).

The service provider is also free to choose any other mechanism (other than the methods define above) to synchronize, provided that it does not require human intervention.

CHAPTER 10

▼▼▼

A DAY IN THE LIFE OF AN M-ES

10.1 INTRODUCTION

This chapter details the message exchanges between an M-ES and the CDPD network during different phases in the operation of an M-ES. An illustration of the different phases of message exchanges an M-ES goes through, from the time it is powered up to the time it exchanges meaningful wireless data associated with the function it is trying to perform helps to visualize the importance of the different phases and provides an overall view as to how to identify and debug operational problems.

This serves as a summary, in a time sequence related to an M-ES, of many of the distinct procedural component parts of the CDPD network covered in previous chapters.

10.2 M-ES CONFIGURATION

The CDPD network is configured with parameter values that have been derived for acceptable performance in a differing array of operational scenarios. The M-ESs, however, can be configured in a manner that provides peak performance for a given type of application. The following sections start with a brief description of important parameters that may have significant impact on the performance in a given application environment.

The M-ESs typically provide several parameters to be configured [1–3]. Clearly, in an embedded application, these parameters are set within the application to be used during the initialization of the CDPD modem. In many modems, the configuration parameters are factory set and are stored in nonvolatile memory and need to be changed only under specialized circumstances. A set of parameters that may have to be reconfigured in a field application is listed next.

- *Channel preference.* CDPD modems will have a configuration option that indicates the preference of choosing A side carrier or B side carrier for CDPD use. This parameter has to be set based on the M-ES's chosen service provider and the associated frequency block.
- *Initial channel acquisition.* Modems use proprietary algorithms to make the initial channel acquisition efficient. For example, a Cincinnati Microwave, Inc. (CMI) modem starts at channel 1 and searches sequentially for a strong signal level (configurable—for example, –90 dBw), drops its search threshold by 5-dB (configurable) increments until a valid CDPD channel is found. Search times can be optimized for specific applications by setting the appropriate modem configuration parameters.
- *Autoregistration.* A modem can be enabled to initiate the registration sequence if it gets reregistered for any reason such as failing to send a reregistration request (ESH packet) before the expiration of the holding timer in its home MDIS. The number of retries may also be another configurable parameter, allowing the modem to retry registration a number of times before discontinuing to register with the network.
- *Roaming restrictions.* In applications where roaming is involved, parameters can be set to place restrictions on the networks the M-ES is allowed to use. Typical restrictions are based on the matching of service provider network identifier (SPNI) and/or local service area identifier (LSAI) of the network with the configured values. Further, a serving network that does not have roaming arrangement with another network where an M-ES is homed will not have the routing set up to communicate with the home MDIS authentication server and, therefore, will not allow the M-ES to register.

The application developer and the user of the modem must be aware of the key configurable settings of the CDPD modem in use so that uncharacteristic behavior of the modem/application can be observed and it can be determined if such behavior can be attributable to the modem configuration. This is especially useful in the initial stages of deployment of a new application/modem combination.

Different schemes are available for an application to interface with the modem, as briefly outlined in the following subsections. Configuration requirements for proper operation in these cases make it necessary that the M-ES user has some level of knowledge in the operation of the TCP/IP and serial line IP (SLIP) protocols.

10.3 INTERFACES FOR MOBILE-END SYSTEMS

Subscriber equipment for CDPD applications are likely to be manufactured in different physical forms and configurations and include, among others, the following:

- Custom devices with embedded modems with preconfigured one/two-way message transfers for applications and installations in vending machines, delivery trucks, and so forth;
- Use of CDPD modems interfaced externally via the serial ports or PC card slots to support applications running in laptop/portable PCs.

Different application-to-modem interfaces are required to support different hardware platforms and application types that would use CDPD for wireless data exchange. Knowledge on these interfaces and an overview of the application execution environment is necessary for efficient troubleshooting of problems that CDPD engineers may encounter in the field. The M-ES can be conveniently thought of as consisting of a mobile application subsystem (MAS)—the computing platform where the application resides—and the subscriber unit (SU)—the CDPD modem, which can be considered to include the subscriber identity module (SIM).

10.3.1 Serial-Interface With AT Command Set

The application residing in the MAS has a serial interface to the SU. The communication between the MAS and the SU are via the published AT command set, which has been extended for CDPD operation. Once the connection is set up, the data transfer between the MAS and the SU are via serial character stream. The basic organization of the this operational configuration is shown in Figure 10.1.

The CDPD modems in this configuration can generally support the following facilities:

- TCP session with a remote unit;
- UDP session with a remote unit;
- Telnet session with a remote unit.

Embedded CDPD applications with burst-mode data transfer requirements are likely to adopt serial AT-mode operation using a TCP session for reliable data delivery requirements and UDP for noncritical and cost-conscious applications. The reduced overhead offered by the UDP protocol and the associated lower cost per data byte transferred would be the only motivating factor for using UDP.

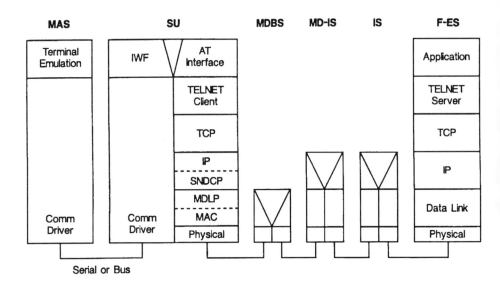

FIGURE 10.1 CDPD AT configuration.

10.3.2 SLIP/PPP Interface

Serial line IP (SLIP) or point-to-point protocol (PPP) mode is the popular MAS-SU configuration when MAS is a laptop or similar device with powerful computing resources. The protocol stack organization in the MAS and the SU for this mode of operation is shown in Figure 10.2.

Most laptop-based CDPD usage will be likely to use the SLIP/PPP interface arrangement.

10.4 REGISTRATION SEQUENCE

Figures 10.3(a) and 10.3(b) illustrate the message sequences exchanged between an M-ES and the CDPD network as the M-ES registers itself with the network. After successfully registering with the network, the M-ES will exchange data with another end system (mobile or fixed) while performing cell transfers (intra-area and interarea) if necessary. At the end of the session, the M-ES may gracefully deregister from the network or may remain registered with the network as dictated by the application requirement. The following distinct phases can be identified in a typical data transfer session of an M-ES using the CDPD network.

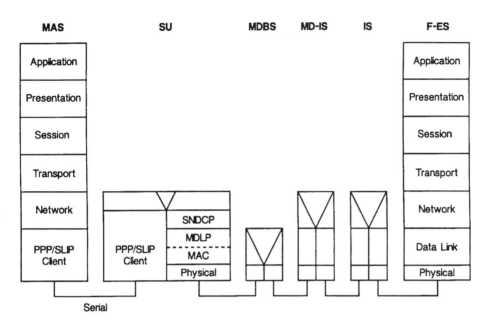

FIGURE 10.2 SLIP Configuration for the MAS-SU interface.

10.4.1 Phase-1: Initial Channel Acquisition

When a properly configured M-ES is powered up, its first task is to acquire a CDPD channel. Assuming there is no past information stored in the M-ES's nonvolatile memory, the M-ES will have to acquire any acceptable CDPD channel. This requires scanning a large number of frequencies, determined by the configured carrier of choice (frequency block A or B), and identifying one that has adequate RSSI level for CDPD operation. The scanning procedure will continue until a CDPD synchronization signal is obtained on a channel stream with a workable RSSI level.

Once the channel is selected and the modem receiver synchronizes to the CDPD forward channel stream, it has to wait for the following two broadcast messages from the MDBS before it can proceed to the TEI assignment phase:

- Channel stream ID message;
- Channel quality parameters message.

Since registration latency will be impacted by the broadcast frequency of these messages, the period for broadcasting these messages is set to less than 5 seconds.

(a)

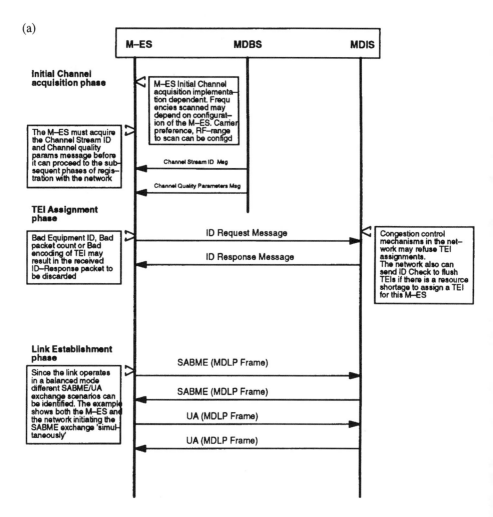

FIGURE 10.3 (a) M-ES power-up link establishment phases, and (b) M-ES key exchange and registration phases.

10.4.2   Phase-2: TEI Assignment

If the SPNI received in the broadcast message is acceptable to the M-ES, it proceeds to send the ID Request message to the MD-IS. The ID Request initiates a procedure in the network to assign a unique identifier, labeled temporary equipment identifier (TEI), to the M-ES so that the MDLP layer frames can be tagged with the TEI in the Address field, enabling a point-to-point link operation. This also allows multiplexing of multiple point-to-point sessions on the same RF channels. The CDPD MD-IS responds with an ID Response message providing the TEI value for the equipment iden-

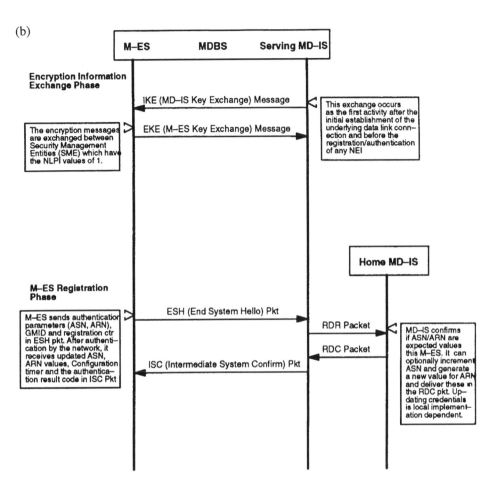

(b)

FIGURE 10.3 (continued)

tifier (a predefined ID that uniquely identifies the modem hardware) sent in the ID Request message.

MDLP parameters are "negotiated" as part of the above messages. In addition, a sleep mode and V.42 bis compression configuration are requested by the M-ES as part of the ID Request message.

10.4.3 Phase-3: Link Establishment

The completion of the TEI assignment phase enables the MDLP layers to establish the link by executing a SABME/UA exchange. For balanced MDLP mode of operation, based on the relative timing from completing TEI assignment, different exchange scenarios are possible. Successful exchange of SABME and UA results in a point-to-point MDLP session being established, and the CDPD network proceeds to the next phase in defining the encryption parameters.

10.4.4 Phase-4: Encryption Message Exchange

The IKE (MD-IS key exchange) message is the first message sent by the network after the MDLP link is established. The security management entity (SME) in the serving MDIS sends this message with the following parameters for encryption:

- base a (public);
- modulus p (public);
- $a^y \text{mod}(p)$ (y private).

The M-ES responds with the EKE (M-ES key exchange) message that contains the parameter, $a^x \text{mod}(p)$ (x private), enabling encrypted message exchanges between the M-ES and the MD-IS.

10.4.5 Phase-5: CDPD Network Registration

Network registration is the key last phase before the M-ES can exchange wireless data with another end system reachable via the CDPD network. Three different registration scenarios can be identified:

- The serving MDIS is also the M-ES's home MDIS. In this case, the RDR/RDC packet exchange is internal to the MD-IS.
- The serving MDIS and the home MDIS are different but belong to the same carrier (SPNI) and the same *routing area*. In this case, RDR/RDC packets flow out/in of the serving MDIS, but may be limited to the same router.
- The serving MDIS and the home MDIS belong to different carriers (SPNIs). Simple interdomain routing facility is employed.

The M-ES will send the ESH packet with its credentials and the serving MDIS will route it to the home MDIS as an RDR packet for authentication. The home MDIS will use an authentication server that will be configured with subscriber profiles and so forth to authenticate the request. The result code, updated credential, and configuration timer will be sent back to the serving MDIS as part of an RDC packet, which is then reformatted as the ISC packet and sent to the M-ES as a response to its registration request.

Successful registration indicates that the CDPD network now has the reachability information for that M-ES and the network is willing to provide data services to that M-ES. Data exchange activity may now be carried out.

10.5 DATA TRANSFER

For a registered M-ES, the CDPD network provides routing of data packets at the network layer. Although availability of IP and CLNP network layer protocol support are defined as part of the specification, current implementations use mainly IP as the network layer.

10.5.1 Burst-Mode Applications

CDPD has been argued to be cost-effective predominantly in applications that require the transfer of a small amount of information (bursty) at regular intervals. Embedded applications where a large amount of CDPD units are deployed, such as monitoring gas pipelines for alarms and vehicle fleet location/dispatch, fit into this category.

In fielded burst-mode applications using CDPD, we can observe the following general operational characteristics:

- Throughput is not a critical factor in application design for burst mode, but worst case latency may be important. The number of M-ESs used in burst-mode applications will be large. These two factors make half-duplex modems, which are inexpensive and have a throughput performance less than that of their full-duplex counterparts, attractive for burst-mode applications.
- Power conservation in the CDPD modems for battery-operated sites may become an important consideration. It is, therefore, likely that sleep-mode operation will be utilized for the burst application types. The sleep-mode operation is requested during the TEI assignment phase when the M-ES can set the required sleep timer (T203) value in the ID Request message. A zero value implies no sleep-mode operation. The MD-IS, based on its implementation, can modify this value. Advanced network implementations should allow the M-ESs to set the value provided it satisfies the low-bound constraint in that T203 should be greater than $2 \cdot$T200. This would facilitate the M-ES application to set the T203 to a value to optimize the operation of the particular application.
- Sleep-mode-enabled applications with moving M-ESs, or in markets with sites using shared frequencies, must be thoroughly tested for operational conformity of the sleep-mode procedures. The M-ES's procedures are critical for proper operational continuity. Occurrence of planned/forced hops or movement of the M-ES to a new physical location may have rendered the last-used RF channel no longer the best channel as the M-ESs do not perform adjacent cell scans for best server selection while in the sleep mode. When the T204 expiration initiates the wake-up procedure, the M-ES simply has to check whether the previous channel that was in use has an acceptable RSSI value (does not have to be the best channel). Otherwise, it will initiate a channel acquisition procedure with the current cell as the best candidate.

- Applications are likely to use the UDP protocol to minimize billable byte count if a connection-oriented session with protocol-level acknowledgment is unnecessary. However, applications such as credit card verification using CDPD will employ a simple TCP-based application to ensure that the query/response is completed with TCP-level protocol integrity.

Figure 10.4 shows a sample of message exchanges in a typical credit card authorization application. The overhead in using a TCP-based exchange is clear. However, the CDPD cost per transaction is likely to be less than 50% of the conventional dial-up authorization facility. Compared to the response latency in a wireline-based connection to the authorization server, the CDPD response will be considerably faster (3–4 seconds compared to 7–12 seconds) since no dialing is necessary.

CDPD infrastructure providers are looking forward to high growth in these types of applications to produce the revenue to justify their investment in CDPD.

10.5.2 Continuous-Mode Applications

All applications that do not fit into the burst-mode category can be assumed to have phases where there is a continuous flow of data in the reverse or forward directions. This application category is likely to use a generic computing platform, such as a laptop, and will mirror applications that are normally performed using a wireline modem. Examples include:

- Business travelers wanting to access their corporate or home networks to query information or perform e-mail/calendar-related functions.
- Applications that require large file transfers, such as banking and warehousing applications, and also for personal use to upload or download files.
- Internet access. The Internet resource is likely to entice the wireless users to increasingly resort to using web browsers to access current information (news, weather, etc.).
- M-ES functioning as a remote X-terminal with X protocol carried in the airlink.

In the above example applications, except in large file transfers, the data flows may not be continuous over a long period of time. However, these applications may involve significant periods, such as downloading a page in the Internet application, where data flow is continuous. Hence, they are loosely categorized as continuous-mode applications.

Users, if given the perception of continuous availability of the 19.2-Kbps bandwidth, which is roughly equivalent to the 14.4-Kbps modems that are in wide use, are likely to find the wireless performance quite acceptable, especially during the early stages of CDPD deployment when the number of concurrent users on a cell or a chan-

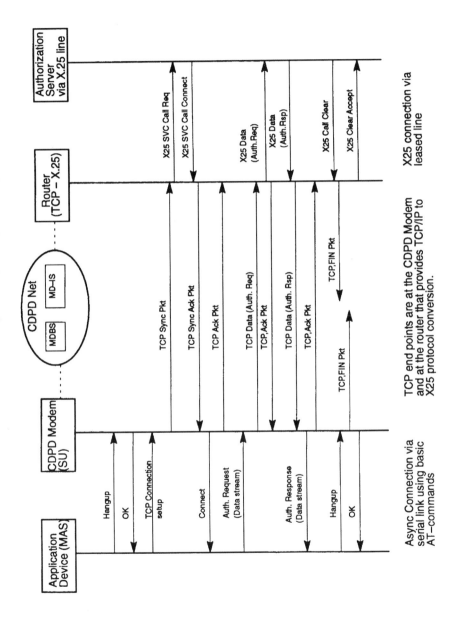

FIGURE 10.4 Message exchanges for credit card authorization.

nel is unlikely to be high. The challenge for the CDPD network service providers is to encourage all possible market sectors with attractive pricing strategies and to continue to provide the channel capacity the users demand.

Since data transfers are continuous for minutes at a time, the following characteristics of the network and the modem will impact the performance of the application. Further, any excessive hang-ups during the application execution also may be attributable to inefficiency in the implementation of any one of the following:

- If a moving M-ES is involved, the efficiency of the cell transfers will impact the throughput. The latency during a cell transfer depends on the following:

  1. Efficiency in the acquisition of the new server and the new channel;
  2. The broadcast interval of the channel stream ID and the channel quality parameter messages from the MDBS;
  3. The resumption of an MDLP session in the newly acquired channel stream by the M-ES by sending a polled RR to the MDBS. Several M-ESs during the early stages of CDPD did not execute this function efficiently, creating significant field problems during file transfers.

- If shared channels are employed in the CDPD network, the continuous availability of unused frequency bandwidth is necessary for successful completion of large data transfers. If the channel availability is absent for more than about 15 sec (the MDLP level T200·N200 value), the MDLP session will be reset and re-establishment will entail TCP-level retransmissions to recover from the link reset. Longer periods of channel unavailability will have further impact on the TCP layer in its calculation of round-trip delay and its exponential adjustment of retransmission timers [4–6].

- If there are other users competing for the limited available bandwidth in the geographic area they are operating, unless the service provider has taken steps to configure the network for *load balancing* and the reverse channel access parameters are configured to allow fair access to the channel (maximum blocks in a burst, minimum time in idle state, maximum retransmissions), customer complaints are likely to be heard.

If the service provider's CDPD network is expected to support a significant number of bandwidth-hogging users, then it is necessary to consider in advance the possibility of using multiple channel streams to distribute the users and using a simple load balancing strategy. With networks that predominantly use dedicated channels, this requirement may be difficult to meet since it may be hard to obtain additional frequencies from the voice system. Deployment of CDMA with the analog voice may free part of the bandwidth, making additional channels available for CDPD (the impact of this is yet to be seen). With networks that already are operating with shared channels, configuring additional traffic channels is relatively straightforward. The key question, even in these cases, is whether there is enough spare capacity in the voice system.

10.6 APPLICATION ISSUES

This section provides insight into the reasons why we may encounter failures in the registration of an M-ES in a CDPD network. Configuration defects, faulty communication infrastructure, possible network congestion, and inadequate RF coverage can be identified as the major causes, as detailed in the following sections. Troubleshooting will involve using the network management features to isolate the cause of the TEI assignment or registration failure and to take remedial action.

10.6.1 Reasons for TEI Assignment Failures and TEI Removals

In normal operating conditions where the M-ES is operating within a strong RSSI coverage area and the supporting CDPD network has no exception conditions (congestion due to excessive traffic or the number of M-ESs has reached a high-level threshold), the TEI assignment phase should complete successfully. However, TEI assignment can fail due to exception conditions in the user side (M-ES) or the network side (MD-IS).

The M-ES will discard the received ID Response message for one of the following reasons:

- The equipment identifier (EID) parameter in the ID Assign message is not equal to its own.
- The expected number of octets is not present in the ID Assign message, indicating a defect in the MD-IS processing or MD-IS to M-ES frame delivery.
- The TEI value is not properly encoded.

Since the ID Assign message has a broadcast TEI (= 0) as the address, all M-ESs are supposed to receive and process it. Therefore, discarding of the ID Assign message does not necessarily mean failure of TEI assignment procedure for an M-ES that discards the ID Assign frame destined for another M-ES.

If the EID parameter in the ID Assign frame matches its own ID, the following conditions may result in TEI assignment failure:

- The version octet does not match the version of the M-ES's MDLP layer.
- The expected number of parameter octets is not present.
- The data link parameters specified are not acceptable.

In these cases, the error recovery is implementation-dependent, but the M-ESs have no direct means of removing the TEI from the MD-IS. The unused TEIs will have to be removed by the audit procedures carried out by the MD-IS.

In addition, the data link layer management entity in the M-ES removes the assigned TEI and notifies the upper layers under the following conditions:

- On detection of persistent loss of channel stream;
- On receiving an error from the data link layer, indicating that an existing data link connection was removed or that it is unable to establish a connection;
- On receiving an error from the data link layer, indicating that it has assumed possible multiple assignment of a TEI value;
- On receipt of an ID Assign message containing a TEI value that is already in use within the M-ES but contains an EID value that is different from its own.

The M-ES removes the TEI also on interarea transfers and when an ID Remove message is received from the network.

The MD-IS discards an ID Assignment request if the version octet does not match the MDLP version or the correct number of bytes is not found in the ID Request message. Audit and recovery procedures in the MD-IS may result in the MD-IS sending an ID Remove message to the M-ES. Typical examples of audit/recovery procedures in the MD-IS that may result in TEI removal are as follows:

- If available resources are exhausted, or TEI information is not available, the MD-IS initiates ID Check requests to selected TEIs. Failure to respond may result in ID Remove messages sent on the failed TEIs so that they are removed.
- An ID Check procedure may also be initiated on TEIs that have not exchanged information for a long period (larger than the configuration timer). This may be followed by an ID Remove to the nonresponding M-ES.
- An implementation-dependent check may be in place to monitor the TEI assignment event to the registration event for the M-ES. Noting that multiple registrations (and NEIs) are possible for a single TEI, TEIs without any registrations for a period may be flushed from the system.

10.6.2 Reason for Registration Failures

Invariably, failure in registration is the result of authentication failure by the authentication server utilized by the home MD-IS of the registering M-ES. If the M-ES is registering in an area other than its own home area, then inter-MDIS communication using CLNP protocol becomes a component of the registration process. Possible reasons for registration failure are:

- The M-ES may not have been provisioned within the CDPD system. The authentication server in the carrier's CDPD network has to have the M-ES provisioned with initial ASN ARN numbers set to zero for initial (first) authentication to be successful.

- The ASN, ARN numbers sent by the M-ES should match the ASN, ARN numbers held by the authentication server of the home MD-IS. If the latest pair does not match, the authentication server attempts a match with the previous ASN, ARN pair for a match (which it should keep). If they, too, don't match, the authentication fails and the M-ES is informed of this as part of the result code sent in the ISC packet.
- The MD-ISs are distributed with home domain directory (HDD), which contains the subnet addresses and the corresponding CLNP address of the home MD-IS. The entries in this directory, defined as part of the authentication server configuration for a carrier's network, should contain the subnet addresses of all the M-ESs that are homed on the carrier's different MD-ISs and also for all other possible visiting M-ESs, which have different CDPD networks defined as their home. Automatic intercarrier distribution of the HDDs is not in place yet. If HDD entries do not include an M-ESs NEI (with mask applied), then the serving MD-IS will not be able to route the registration request to the home MD-IS for authentication and the registration will fail.
- If a visiting M-ES is involved, the inter-MDIS communication mechanism must be in place. The registration will fail if the serving-MDIS cannot communicate to the home MDIS.

  This will involve the intermediate system (router) connected to the serving MDIS have the CLNP IS-IS Level-1 routing protocol enabled at that port, and the CLNP IS-IS Level-2 routing enabled. The router at the home MDIS facility should also have a similar configuration. Interdomain communications and the associated additional routing protocols may be necessary when domain-level networks are brought into CDPD use.

Once the registration is successful, the reachability information of that M-ES is known to the network. Registration completion also signifies that the network has expressed its willingness to provide data transfer services to that M-ES.

10.6.3 Reason for Loss of Registration

Operationally, an M-ES, once registered with a CDPD network, should maintain the registration for an indefinite period of time even with minimal (or no) data transfers provided some basic procedures are followed. This section lists possible reason why an M-ES may lose registration:

- The home MDIS maintains a holding timer (currently, the default is 8 hours), which is first started when the initial ESH packet for an M-ES is received. If a reregistration request from the same M-ES did not arrive before this time expires, the registration/location information is removed from the system.

- If the M-ES has requested sleep-mode operation and the wake-up procedure initiated by the network fails, the network will flush the NEI/TEI from the system.
- If for any reason the TEI supporting the registration of an NEI(s) is removed, the NEI registrations will be flushed. During an interarea transfer, the TEI and the registration will be removed as part of the normal operation.

M-ES configured for autoregistration will initiate a new session once it recognizes that the TEI is removed.

10.7 SUMMARY

The step-by-step illustration of the interaction of a CDPD modem with the CDPD network contains information that is valuable to both users of CDPD applications and the network service providers. The idiosyncrasies of different modem types, especially the configuration aspects, need to be handled patiently by the users of the modems. With an increasing number of third-party vendors manufacturing CDPD modems, the configuration/initialization aspects of the modems are likely to improve with time. Further, only users who are using the CDPD modems for ftps, simple terminal emulation, or setting up continuous pings to an end system (M-ES or F-ES) need to handle the low-level aspects of modem configuration. For "embedded" CDPD applications, the application itself will likely do all the configuration necessary, removing this burden from the users.

Note: All CDPD modem user manuals contain the AT command set.

References

[1]   Cincinnati Microwave, Inc. (CMI), Micro Dart CDPD modem user manual.

[2]   PCSI Ubiquity CDPD Modem user manual.

[3]   Sierra Wireless CDPD Modem user manual.

[4]   RFC-793: Postel, J., Transmission Control Protocol, September 1981.

[5]   Karn, P., and P. Partridge, "Improving Round Trip Time Estimates in Reliable Transport Protocols," *Proceedings of ACM SIGCOMM'87*, Stowe, VT, 1987, pp. 2–7.

[6]   Jacobson, V., "Congestion Avoidance and Control," *Proceedings of ACM SIGCOMM '88*, Stanford, CA, 1988, pp. 314–329.

CHAPTER 11
▼▼▼

CDPD DEPLOYMENT AND OPERATIONAL ISSUES

11.1 INTRODUCTION

The CDPD deployment issues faced by network service providers, and the M-ES/application operational issues faced by customers who use the CDPD network with different application categories employing CDPD modems from different manufacturers, are identified and detailed in this chapter. Evolving issues in the CDPD technology, such as equal access provision and interoperability, are treated in the Chapter 12.

11.2 CDPD OVERLAY ARCHITECTURES

The operational AMPS system architecture places constraints on the type of CDPD configuration that can be deployed as an overlay network. Typically, the carrier's choice is to implement a one-to-one arrangement where a CDPD MDBS, matching the configuration of the AMPS, is installed in each AMPS site. This section outlines the different CDPD configurations that can be found in different markets and the impact of these configurations in sharing of the RF hardware, cell-site commissioning, and general RF engineering.

11.2.1   OMNI Cells

OMNI configuration of a cell site is the simplest in terms of the cell-site setup, but offers the least flexible configuration in tuning the cell site for optimal performance to work in conjunction with its neighbor cells [1–3]. Further, OMNI configurations are not the best candidates to combat co-channel interference. In a typical AMPS network, only 5–10% of the cell sites are configured as OMNI sites.

11.2.2   Multisectored Cells

Use of directional antennas allows a cell to be split into sectors. The three-sector case with seven cell clusters is widely used in the field. This configuration helps to reduce co-channel interference, but requires 21 distinct channel sets and has the disadvantage that there will be fewer channels available per sector. However, the capability to use smaller cells through sectorization outweighs these drawbacks, resulting in a higher capacity system.

The cell site will contain three sets of antennas, one set per sector. Antenna configuration per sector will be similar to that required for the OMNI configuration. As seen earlier, this could involve separate transmit and receive antennas or a combined transmit/receive antenna with a duplexor splitting the transmit and the receive bands inside the cell site hardware. An additional receive antenna may be employed per sector for diversity reception.

11.2.3   Simulcast

Simulcast configuration in the AMPS application is a variation of the sectorized configuration where the same setup (control) channel is broadcast in all the sectors. This organization introduces the operational concerns in the version 1.1 CDPD networks where the AMPS setup channels are defined as the reference channels used in the best server selection procedures. Clearly, in sites using dedicated channels for CDPD, the reference channels can be the CDPD dedicated channels and the simulcast configuration will not have any impact in the operation of the CDPD. The CDPD network will only "see" a sectorized configuration.

11.2.4   Sparse Deployment

If the CDPD network is implemented with exclusively dedicated frequencies and the CDPD usage is limited or being implemented on a trial basis, the carrier has the option to deploy a sparse CDPD network. In this arrangement, the CDPD cells can be designed to cover a larger geographic area than the AMPS counterparts by operating at high transmit power levels at the cell-site. This would allow the service provider to install CDPD equipment at selected AMPS sites, reducing the capital cost required to support CDPD facility.

In addition to reduction in the CDPD traffic capacity, this organization may also require more thorough RF engineering studies if the CDPD coverage is to be adequate. Further, the service provider must have a road-map and the cell deployment procedures worked out to have the CDPD network upgraded to AMPS-type coverage should the initial deployment prove successful and require increased traffic capacity in the CDPD network.

11.3 FREQUENCY ALLOCATION AND MANAGEMENT

In a CDPD configuration that is one to one with the AMPS cellular setup, the RF engineering required for CDPD for frequency allocation, reuse strategy, and transmit power level setting to define cell boundaries will match those performed for the analog voice system. Sectorization and clustering of cells will be the same.

If the availability of channels for CDPD use is scarce, a shared-channel arrangement will be used where the frequency allocation will have already been done for the voice system. In this scenario, a subset of the voice channels can be assigned for CDPD use. Where dedicated channels are available, specific channel(s) have to be reserved for CDPD use in each of the distinct sets of frequency pools defined for the cell cluster types in use.

If the RF engineering methodology adopted by a service provider is not rigorous, problems related to RF coverage, interference within the CDPD spectrum, and interference with the voice system may occur in the fielded CDPD network.

11.3.1 Co-Channel Interference

Reuse of frequencies to maximize spectrum utilization and to increase the cellular system capacity implies that two or more distinct communication channels will be using the same frequency. The interference caused to a specific channel by all the other channels using the same frequency is called co-channel interference [4]. Co-channel sites have to be separated by a minimum distance to avoid providing unacceptable levels of co-channel interference.

Co-channel interference rejection is a measure of the receiver's ability to receive a modulated signal on its assigned channel frequency in the presence of a second modulated signal on the same assigned channel frequency (the interfering signal).

In CDPD, the co-channel interference rejection is measured in block error rate (BER) when the desired signal is subjected to a Rayleigh-faded time-dispersive channel (mixed with the same channel with a delay of 8 ms) and corrupted by an interfering Rayleigh-faded channel modulated with a nonbit-synchronous pseudo-random sequence without time dispersion. The minimum requirement is that when the desired signal RSSI is −85 dBm, the interference signal is −105 dBm, the clock generators of these two signals differ by 25 ppm, and Rayleigh fading is represented by a doppler frequency of 65.31 Hz (80 km/hr simulated), the maximum BER must be less than 15%.

In the equivalent specification in the AMPS system, the cell/RF configuration should provide a co-channel carrier to interference ratio (*CIR*) of 18 dB or greater for 90% of the coverage area. This figure is based on the criterion that at an 18 dB carrier-to-noise ratio (*CNR*), 75% of the users consider voice quality to be good on an FM channel (30 kHz) in a multipath environment.

For a cellular system with cell radius *R*, the co-channel cell site distance *D* can be shown to be

$$D = R \cdot (3N)^{1/2}$$

If we assume that the interference in a cell is predominantly from co-channels cells, the carrier to noise (interference) ratio, for a 7-cell cluster, can be written as

$$C/I = C/(6 \cdot I_{\text{one co-channel cell}}) = 1/6 \cdot (D/R)^4$$

assuming a free space path loss of 40 dB/decade.

This represents the worst case when all the six co-channel cells are using the same frequency as the cell they are causing interference to. To provide a *C/I* ratio of at least 18 dB (AMPS, equivalent to a ratio of 63.1), we can show that the *D/R* ratio has to be greater than 4.4.

Diversity reception—thorough RF engineering analysis in RF channel allocation, use of sectorized antennas, use of tilted antennas to provide a notch effect on the horizontal RSSI pattern, and effective utilization of group color codes—can be used to reduce the impact of co-channel interference in CDPD operation. Increasing the number of cells in a cluster increases the co-channel separation distance with respect to the cell radius, thereby reducing the co-channel interference. However, this limits the traffic capacity supportable by the CDPD network (lower number of RF channels per cell).

11.3.2 Adjacent Channel Interference

Adjacent channel interference is the result of transmissions in the RF channel adjacent to a channel containing energy within the bandwidth allocated to the current channel. In an *n* cell cluster design of the cellular system, by assigning frequencies to cells that are separated by *n* channels, the impact of adjacent channel interference can be minimized.

Adjacent channel *selectivity* is a measure of the receiver's ability to receive a modulated input signal on its assigned channel frequency in the presence of a second modulated input signal spaced either one channel (30 kHz) above or one channel

(30 kHz) below the assigned channel frequency. Selectivity from interference channels that are 60 kHz above or 60 kHz below the assigned channel is called *alternate channel selectivity*.

CDPD specifications for adjacent/alternate interference selectivity require that for a CDPD receiver operating at RSSI levels of 3-dB higher than the receiver minimum sensitivity, a 16-dB higher adjacent channel or a 60-dB higher alternate channel should not produce a BER of greater than 5%.

A CDPD system operating within a well-designed AMPS RF frequency allocation scheme can be expected to have no significant level of adjacent channel interference.

11.3.3 Coordination of AMPS and CDPD Frequency Pools

Since it is likely that the AMPS and the CDPD systems will be managed by nonintegrated network management systems, frequency pool configuration changes will be performed and coordinated manually. In fielded AMPS/CDPD systems, changes to cell configurations resulting from the addition of cell sites or cell splitting occur frequently. RF-channel allocations to cells also change often. It is vital that procedures are in place so that when voice pools change, the associated CDPD frequency pools are also modified appropriately.

11.3.4 Efficient Algorithm Match

Known voice systems usually implement two different algorithms for frequency assignment to the voice radios:

- *Serial assignment*. The frequency pool is scanned from top to bottom and the first available frequency is assigned for voice use.
- *Equal channel use*. The frequencies are selected such that there is equal utilization of the different frequencies.

Service providers generally choose the serial assignment scheme so that if a cell has some of its first tier co-channel cells have the pools defined in the inverse order to that of the current cell, the incidence of co-channel interference can be minimized.

Similar argument can be made for the CDPD cell. If the frequency pool is defined in the reverse order to that of its overlaid AMPS cell, the voice collisions can be minimized.

CDPD implementations can provide the service provider the ability to choose one of several algorithms to dynamically reorder the CDPD channels to minimize collisions. These algorithms must be enabled where possible to reduce the impact of the voice system and thereby increase the throughput of the CDPD traffic channels.

11.4   CELL CAPACITY AND LOAD MANAGEMENT

11.4.1   Multiple MES Operation

The issues related to multiple M-ES operation are the proper setting of MAC-level parameters that allow sharing of the bandwidth in the reverse channel on an equitable basis among the competing M-ESs and whether any other schemes can be implemented within the MDBS/MD-IS to further assist in the proper distributing of forward and reverse channel bandwidths. These relate to a single channel stream and the expectation is that a higher level load-balancing scheme would be forcing a balanced distribution of M-ESs to the active channel streams in a cell.

The digital sense multiple access with collision detection (DSMA/CD) algorithm and the associated MAC-level parameters provide the M-ESs a fair method for equal access to the reverse channel stream. A simple algorithm to dynamically vary the maximum number of blocks allowed in a burst can be implemented based on the number of M-ESs using the channel. This may have some impact at very high levels of continuous utilization of a channel stream.

Forward channel stream behavior is implementation-dependent and also depends on the data network characteristics (including delays) external to the serving MDIS. A "fairness" scheme can be implemented at the MDBS to forward queued packets in a roundrobin scheme rather a FIFO scheme. Pertinent issues related to the forward stream data transfer are as follows:

- The maximum MDLP frame size is 128 bytes. If the IP packet size is much larger (1,024 bytes or more), with a transmit window size of 7 to 15, forward direction packets would tend to be sent in "groups of frames" by the MD-IS.
- The roundrobin scheme at the local MDBS end may in fact distort the overall end-to-end forward path behavior of the different M-ESs.

These issues need careful consideration and testing before an algorithm is implemented for field deployment.

11.4.2   Load Balancing

In a cell that supports multiple CDPD channels, the network must make an attempt to ensure that M-ESs are fairly distributed across the available channel streams. The M-ESs during the channel acquisition procedure are likely to scan the frequency pool broadcasted to them in a sequential mode, thereby choosing a frequency that is above the others (used by that cell) in the frequency pool list. Therefore, if the network does not actively participate in load balancing, the distribution of the M-ESs on the available channel streams in a cell is likely to end up as nonoptimal.

CDPD channel streams provide a complex medium for load balancing, given the nature of the applications that require varying amount of air-bandwidth and the channel access/utilization constraints imposed on the using M-ESs. The following factors must be taken into account when implementing effective load-balancing algorithms. Networks have not matured to the point where multiple channel streams are in widespread use, and therefore the load-balancing issue has not been dealt with adequately by the infrastructure providers.

- M-ESs competing for use of airlink fall into two important groups, which need to be considered separately when load balancing is attempted:
  1. M-ESs that have a significant amount of data to be transferred in the forward direction or reverse direction. These M-ESs are characterized by the fact that, if given the chance, they are capable of occupying the full bandwidth of the channel stream (forward or reverse depending on the data transfer direction) (full-bandwidth M-ESs).
  2. M-ESs that have limited amount of data to be transferred and are not directly affected by throughput figures. These may have a requirement for worst case latency.

- If there are more full-bandwidth M-ESs in operation than there are number of available channel streams, load balancing can achieve 100% utilization of the channel streams by having at least one full-bandwidth M-ES in each channel stream. To ensure fairness in providing bandwidth to the M-ESs, the load-bal ancing algorithm should first attempt to distribute evenly on all available channels the full-bandwidth M-ESs that are currently using the cell. Similarly, bursty M-ESs can also be evenly spread across the channel streams.

Load-balancing algorithms will have to use simple techniques to identify the M-ESs' bandwidth consuming potential to be effective in distributing them in a timely manner across available channels in a cell. In practice, the load balancing may prove to be more difficult to accomplish if we consider the following additional technical issues:

- In a cell where shared channels are employed, load balancing has to be aware of the forced channel hops, after which the M-ESs on the hopped channel need not necessarily reuse the same channel stream.
- Switch channel messages are the primary tool through which a load-balancing algorithm can move an M-ES from one RF channel to another. Switch channel messages are broadcast and therefore reliable delivery is not guaranteed. Further, if half-duplex M-ESs are present, we cannot command them to switch reliably unless a consistency check at the MDBS as to whether a switch has occurred is verified.

11.5   GMSK AND FM RADIO OPERATION IN THE CDPD MODEMS/MDBS

The Gaussian minimum shift keying (GMSK) modulation technique specified for CDPD operation has been described in some detail in Section 4.5, and its core MSK technique has been shown to contain the following properties:

- MSK is a constant envelope modulation scheme suitable for nonlinear and power efficient amplification.
- MSK has phase continuity at bit transitions.
- The receiver can employ coherent or noncoherent detection capability [5].
- The resulting modulated MSK signal is equivalent to frequency shift keying (FSK) with signaling frequencies of $f_c + 0.25/T$ and $f_c - 0.25/T$ (i.e., the frequency deviation is half the bit rate).

The main lobe of the MSK spectrum is roughly 50% wider than that of QPSK signals. The GMSK modulation technique improves the spectral efficiency by employing a premodulation Gaussian lowpass filter (GLPF) without affecting the first three properties of the MSK scheme described above. The BT product is a compromise between having a compact spectrum and obtaining good error rate performance; CDPD uses a value of 0.5 as the BT product of the GLPF.

Since MSK has been shown to be equivalent to a special case of FSK with a modulation index of 0.5 (frequency deviation of half the bit rate), GMSK modulators can be implemented using FM modulators or using quadrature cross-correlated architecture (IQ modulator), as shown in Figure 11.1.

Similarly, GMSK demodulators can be implemented using a noncoherent technique or a coherent technique, as shown in Figure 11.2.

Use of FM modulators and FM demodulators for CDPD modems and the MDBSs are common. With FM modulators and demodulators, hardware cost is low and RF sections can be shared with those for analog voice systems that use FM modulation. However, characteristics intrinsic to FM modulators/demodulators present interoperability issues with hardware that employs IQ modulators and coherent demodulators. The pertinent characteristics of the different modulators/demodulators that lead to interoperability issues are as follows:

- In modulators using the voltage-controlled oscillator (VCO)-FM technology, the modulation index can drift over time and temperature. Further, calibration and adjustment of the modulation index is not trivial with the premodulation filter in place. Long patterns of 1's and 0's must be used to accurately adjust the modulation index to within 5% tolerance. Potentiometers are generally used in conventional FM radios, allowing only coarse adjustment of the modulation index.

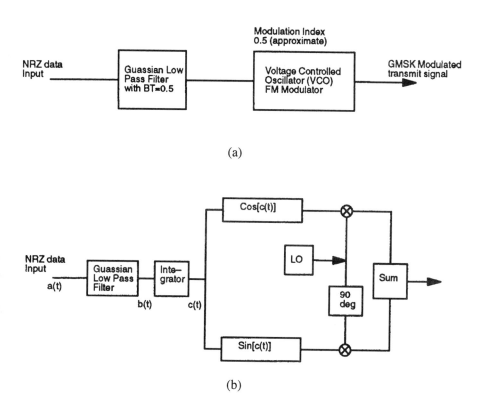

FIGURE 11.1   GMSK modulators: (a) GMSK modulator using FM modulation and (b) IQ modulator.

- For coherent demodulation, the modulation index is extremely critical for providing high receiver performance.
- The intersymbol interference (ISI) level in a noncoherent demodulator (using FM discriminators) does not depend on the modulation index. Therefore, the demodulator performance in a noncoherent detection system is not sensitive to the accuracy of the modulation index.

The above characteristics indicate that when we employ a VCO-FM–based CDPD modulator transmission and a coherent demodulation scheme, unless the VCO-FM modulator is accurately calibrated to provide a stable modulation index of 0.5, we may encounter problems in reception. Extra effort needs to be applied to the

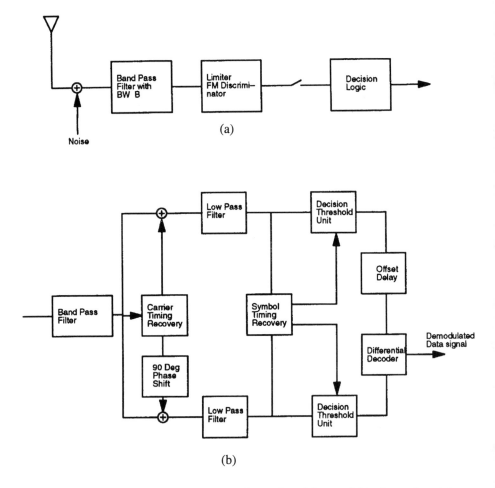

FIGURE 11.2 GMSK demodulators: (a) noncoherent demodulator and (b) coherent demodulator.

transmitter modulation hardware to enforce that it complies with the CDPD specifications related to modulation index.

11.6   HALF-DUPLEX MESs

Half-duplex M-ESs have optimized RF circuitry that allows only reception or transmission at any one time. These M-ESs are cheaper and perform adequately in a burst-mode application. For applications that require transfer of large amounts of data, the decreased throughput performance of a half-duplex M-ES may be a drawback.

While transmitting information on the reverse channel, the half-duplex modems cannot listen to the forward channel. Any data destined to the M-ES in the forward channel during this time window will be lost and will have to be recovered by MDLP layer retransmission procedures. Increased number of selective reject (SREJ) frames at the MDLP level sent by the CDPD modem (in the reverse direction) can be noticed in a half-duplex M-ES session using the CDPD network. To minimize the effect of data loss from the forward channel, the half-duplex M-ESs are restricted to sending only one block in a burst. These two factors account for the decreased throughput performance of half-duplex M-ESs.

11.7 ANTENNA ISSUES

RF coverage issues for CDPD are similar to those of the AMPS system it is overlaid on. Existing antenna arrangement supporting the AMPS system is invariably used to support the CDPD system. The integration of the CDPD systems and the AMPS systems is done at the RF level external to the power amplifiers for transmission. The combined received RF signal is split and fed to the CDPD receive inputs.

11.7.1 Cell-Site Antennas

In a cell site, usually a standard 6-dB or 9-dB transmit antenna is used [2,6]. If the number of transmitting radios (AMPS) is large (example 32), and if the combiners are limited to 16 channels, then multiple transmit antennas may have to be used. One or two receive antennas can be used if diversity is employed. Further, a transmit antenna may also serve as a receive antenna if duplexors are employed to separate the transmit and receive signals. A sectorized cell requires one set of antennas per sector.

11.7.2 Mobile Antennas

Common mobile antennas used in the field (cellular and CDPD) are shown in Figure 11.3 and briefly described next.

A quarter-wave antenna is about 3.5 inches long and has a unity gain. It has been shown to perform well in densely populated areas where buildings and terrain block the line of site between the cell-site antenna and the mobile. A collinear design with a 1/2 upper radiator and a 1/4 lower radiator separated by an air wound inductive coil produces a 3-dB gain antenna about 14 inches long. Similarly, two 1/2 lengths and a single 1/4 length separated by two inductive coils provides a 24-inch long 5-dB gain antenna.

Choice of these different antennas can increase the performance of an M-ES in areas where the RF coverage is not sufficient for low-gain antennas to work.

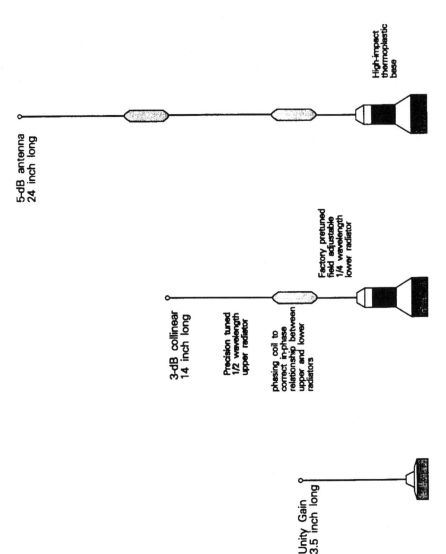

FIGURE 11.3   Mobile antennas.

11.7.3 Typical Problems With Cell-Site Antennas

11.7.3.1 Dead Spot

Higher antenna gains are achieved by compressing the main beam width produced by a half-dipole omnidirectional antenna (with a circular pattern for the main lobe) in the vertical plane. Minor lobes are an unfortunate side product of these higher gain antennas. Associated with the minor lobes are nulls where in identified angles there is no radiated power. Therefore, although the distant coverage is better in a high-gain antenna, dead spots can occur in an area where coverage is desired.

Lowering of antenna height, reducing the antenna gain, or electrically tilting the antenna are ways by which the null areas can be moved. The impact of any one of these in the whole coverage area needs to be considered before a choice is made.

11.7.3.2 Transmit/Receive Antenna Isolation

It has been known that a reduction in receive sensitivity is caused by inadequate horizontal/vertical spacing between the receive and transmit antennas. The specific reasons are as follows:

- The receive in-band noise increase is caused by a cosite transmitter.
- Strong off-channel signal causes a gain reduction in the receiver low-noise amplifier.

Adequate antenna spacing and filtering of the out-of-band transmitter channel noise by devices such as receiver multicouplers and duplexors will help to prevent or alleviate the receiver desensitization.

11.8 SYSTEM TUNING

This section provides an overview of the parameters that can be changed by the CDPD network operator to affect the operation of the network. Many of the parameters are usually set at the default values recommended in the specifications, while some parameters, as identified in the following sections, will require adjustments for optimal operation of the network. In addition to these parameters, the infrastructure implementors will have their own system specific parameters that may require adjustments. These include parameters related to statistics collection and so forth, and are not covered in this section.

11.8.1   MAC Layer Parameters

Configurable MAC-layer parameters fall into two groups: those that influence channel access for multiple M-ES operation and those that govern MAC-layer error recovery. In the field, these parameters rarely need to be changed once a default setting is chosen during installation of the network.

Channel Access Parameters

The parameters that govern the access procedures to the reverse channel and their recommended default values are as follows:

- Maximum contiguous blocks in a burst (64 blocks);
- Maximum entrance delay (35 microslots);
- Minimum time in idle state (8 microslots).

With each block representing approximately 35 bytes, the value for default maximum contiguous blocks represents 2.25 KB of data. The default maximum entrance delay translates to 110 ms and the default minimum time in idle state translates to 25 ms. The relatively large default value for the maximum contiguous blocks (64), in particular, may favor M-ESs with powerful processing hardware, capable of assembling multiple MDLP frames (128 bytes maximum) into a single burst. The powerful M-ESs are then bound to obtain disproportionately large fractions of channel bandwidth, mainly in the presence of half-duplex M-ESs sharing the channel. A value of 32 or 16 may be chosen based on the characteristics of the M-ESs using a channel.

Implementors may decide to have different sets of values for maximum blocks and minimum time in idle state, to be used dynamically based on the number of M-ESs actively using a channel and the reverse throughput in the channel stream.

Error-Recovery Parameters

Error recovery due to reverse channel collisions or due to weak received signal strengths at the MDBS involves the following parameters at the M-ES end of the MAC layer:

- Minimum count for backoff idle interval (4);
- Maximum count for backoff idle interval (8);
- Maximum transmit attempts (13–15).

The backoff parameters are rarely changed from the suggested defaults. The maximum transmit attempts is a useful parameter to experiment with if one observes an increasing number of channel scans due to channel exception triggered by exceeding retransmission attempts. If one suspects weaker RSSI values or a large number of M-ESs using the same channel, this parameter can be increased to 18–20 to see if an improvement is obtained. The principle here is that the recovery at the MAC level should happen within the time window allowed by the layer above it, the MDLP layer (T200 × N200).

11.8.2 MDLP Layer Parameters

The MDLP parameters fall into three groups. As for the MAC layer, except in exceptional circumstances it is not necessary to change the MDLP parameters from the recommended default values.

TEI Management

Configurable parameters associated with TEI management are as follows:

- T202 – ID request timer (5 seconds);
- N202 – ID request retransmissions (3);
- T201 – ID check timer (5 seconds).

The above default values will support all TEI management scenarios and are likely to stay unchanged.

Sleep Mode

Configurable parameters associated with sleep management are as follows:

- T203 – Element inactivity timer (30 seconds);
- T204 – TEI notification timer (60 seconds);
- N204 – TEI notification retransmissions (5).

The element inactivity timer can be increased if 30 seconds is shown to be impacting the performances of the M-ESs whose operational characteristics force them to enter and exit the sleep mode too frequently.

MDLP

Configurable parameters associated with MDLP protocol are as follows:

- M-ES: T200—Retransmission timer (3 seconds);
- M-ES: N200—Retransmission counter (3);

- MD-IS: T200—Retransmission timer (5 seconds);
- MD-IS: N200—Retransmission counter (3);
- K—Window size (15 for Tx and Rx).

Large window sizes will require the availability of additional memory resources in the MD-IS. The T200 or N200 can be increased slightly if link resetting becomes an issue, but the increase should be cognizant of the TCP timeout values.

11.8.3 Radio Resource Management Parameters

The parameters that govern the radio resource management operation would likely need modifications in the field to obtain optimal operating configurations.

RRM Message Broadcast Intervals

The different messages and the default broadcast intervals are shown below:

- Channel stream ID (3 seconds);
- Channel quality parameters (5 seconds);
- Channel access parameters (60 seconds);
- Cell configuration (60 seconds, distributed).

The channel stream ID message impacts the latency during an M-ES handoff. The M-ES cannot use the newly acquired channel until it has received the channel stream ID and the channel quality messages from the MDBS on the new channel stream. The frequency at which these two messages are broadcast must be less than 5 seconds to achieve acceptable channel hop latencies.

Cell configuration message interval must be less than the maximum channel scan time. This is to ensure that the M-ES receives cell configuration messages for all the neighbor cells before it selects the best server channel for a scan triggered by the expiration of scan time.

Channel Hop/Handoff Parameters

The parameters that impact handoff will require close tuning for each cell site to obtain the maximum performance from the network. Terrain conditions, the closeness of neighbor cells, and the concentration of M-ES traffic in the geographic area around the cell with respect to its neighbors will influence the setting of the following parameters:

- RSSI hysterisis (8 dB);
- Scan time (90 seconds);
- Scan delta (8 dB);

- RSSI average time (5 seconds);
- BLER threshold (20 %);
- BLER average time (5 seconds);
- RSSI bias (0 dB).

Since the 1.1 specification requires "continuous" selection of the best channel, the block error rate (BLER) threshold will be typically set to a higher value than those used on the 1.0-compatible networks. A positive value of RSSI bias (defined for each neighbor) will artificially move the neighbor's boundary towards the center of the current cell. This can be used to move M-ESs geographically located within the current cell to the neighbor cell. A negative RSSI bias will have the opposite effect—it will shrink the boundary of the neighbor towards its own center. Hysterisis and scan delta values may require minor adjustments based on the radio coverage and the terrain-related coverage variability. The RSSI hysterisis value setting is a balance between flipflopping sensitivity and the dragging of cell boundary. Similarly, the scan delta value is a balance between frequency of scans and delay in the choice of the best server channel.

11.9 CURRENT STATE OF CDPD DEPLOYMENT

The CDPD Forum regularly publishes the current state of CDPD deployment, categorized as, full cellular (coverage equivalent to cellular voice), core (existence of significant coverage), and initial deployment (one or more cell sites available for customers).

The service areas are also grouped as metropolitan statistical areas (MSAs) and rural service areas (RSAs). The MSA denotes one of the 306 largest urban population markets and RSA denotes one of the 428 less populated areas of the country as designated by the U.S. Government. The report card of October 11, 1995, detailing the state of CDPD deployment is shown in Appendix B.

The CDPD Forum has a web site that can be accessed by pointing the browser to http://www.cdpd.org. The latest information compiled and distributed by the CDPD Forum can be obtained via this web site.

References

[1] Lee, W. C. Y., *Mobile Communications Design Fundamentals*, Wiley Interscience, 1993.

[2] Mehrotra, A., *Cellular Radio Performance Engineering*, Norwood, MA: Artech House, 1994.

[3] Mehrotra, A., *Cellular Radio: Analog and Digital Systems*, Norwood, MA: Artech House, 1994.

[4] Macario, R. C. V., *Cellular Radio: Principles and Design*, McGraw Hill, 1993.

[5] Feher, K., *Wireless Digital Communications*, Englewood Cliffs, NJ: Prentice Hall, 1995.

[6] Balanis, C. A., "Antenna Theory: A Review," *Proceedings of the IEEE*, Vol. 80, No. 1, Jan. 1992, pp. 7–23.

CHAPTER 12
▼▼▼

EVOLVING CDPD ISSUES
OF THE FUTURE

12.1 ADAPTATION TO EVOLVING CELLULAR
TELEPHONY TECHNOLOGY

A common core infrastructure for CDPD is being built, and the CDPD footprint will become ubiquitous across the country (and maybe even Canada and Mexico) in the next couple of years. One of the many characteristics of CDPD is the use of airlink resources of the advanced mobile phone system (AMPS) infrastructure. However, the CDPD network is an evolvable network that can interwork with alternative RF technologies.

As newer cellular telephony technology evolves, it would be useful to leverage the existing CDPD solution to support wireless data over these technologies. The adoption of alternative RF technologies can be accommodated with minimal disruption to the system architecture.

Figure 12.1 depicts the network architecture across the airlink interface. The airlink consists of the protocols that directly interface to the M-ES. The "other side" of these protocols interface to the MDBS or the MD-IS.

As we work through this architecture, we come to the following conclusions:

- The IP/CLNP network layer protocols are widely supported and thus allow ease of application attachment. Support for these network protocols should be

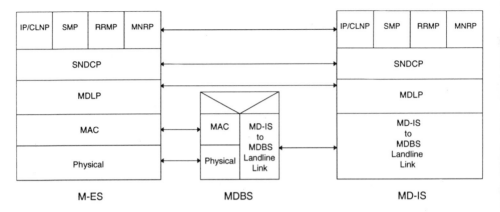

FIGURE 12.1 Network architecture across the airlink.

maintained with alternative RF technologies. In addition, support for emerging network layer protocols like IPv6 will provide the much needed relief from the persistent scarcity of network addresses. With the advent of paging class devices performing two-way messaging over CDPD or CDPD-derived wireless data networks, IP addresses just won't be adequate in number to satisfy the numbers predicted for this class of devices.

- The mobility management solution consisting of SNDCP, security management, authentication procedures, and key exchange procedures can be easily adapted and reused over the emerging alternative RF and cellular telephony technologies.

- The MDLP protocol provides a reasonably efficient link layer protocol that includes support of mobile system movement, mobile system power conservation, RF channel resources sharing, and error recovery. This layer can be maintained with very little change when evolving a wireless data solution over alternative RF technologies.

- The RRM provides control and management over the mobile device's use of the RF resources. The protocol as it exists for CDPD today is specific to an overlay network over AMPS RF infrastructure. Similar functionality will be needed in an alternate set of protocol functions to alternative RF technologies.

- The MAC and physical layer protocols will most likely require appropriate replacements based on the specific RF technology.

In IS-136, the latest TDMA standards that supercedes IS-54, provisions for digital data transfer are incorporated as defined in IS-130 and IS-135. These standards, which cover digital asynchronous data transfer and group 3 fax, make possible error-free file transfers at 9,600 bps (uncompressed). In the short term, if TDMA infrastructure providers implement these features, TDMA may become a serious competitor to the CDPD network.

12.2   IMPACT OF CDPD ON PCS

PCS services that are expected to become popular fall under the two broad areas of voice and data. While CDPD is not a very elegant solution to providing a transport mechanism for voice applications, the data service lends itself to the use of a CDPD-like paradigm. The richness of CDPD combined with the addition of some paging device-oriented features, and also the elimination of unnecessary features like channel hopping, can provide maximized use of the available technology.

Narrowband personal communications services, in particular, can make use of technology rooted in CDPD. These services are typically useful for short messaging applications like two-way paging, e-mail, telemetry, personal home security, electronic calendar updates, tracking, and dispatch-oriented applications.

AT&T Wireless Services has already announced personal Air Communications Technology (pACT), a new open standards-based architecture for two-way messaging in the narrowband PCS frequencies (the 900-MHz band). The forward and reverse 12.5-kHz channels are in the range of RF channels 1 to 39 assigned for transmit and receive within the 500-kHz band allocated for the narrowband PCS services. The pACT architecture has been derived from a base CDPD technology and retains the features of IP connectivity, data security, authentication, and encryption that are an integral part of CDPD.

The pACT architecture is symmetrical, with inbound (forward) and outbound (reverse) data rates being equal in capacity, and therefore speed. A network using pACT will be configured like a cellular network, comprised of intelligent base stations that are equipped with transmitters and receivers to route messages. The mobility management functions of CDPD are made use of to provide seamless mobility from one cell area to another. The paging network is always aware of the current location (in terms of cell and sector) of the paging device. This paradigm of paging is different from the more conventional acknowledgment-based paging systems in use where the paging network transmits the page in the home area first. It then awaits an acknowledgment from the paging device. If not answered in that zone, the paging signal is broadcast repeatedly over a larger area.

12.3   CIRCUIT-SWITCHED CDPD

Truly ubiquitous CDPD service deployment across the country will probably take the next couple of years. The service providers have to invest money into infrastructure equipment and the management of it. This investment was started in 1994, but the present pace of service deployment has slowed down as the subscriber market has been not very responsive, mainly because a lot of applications that can exploit CDPD are still in the making.

The existing CDPD technology can be complemented by the circuit-switched CDPD (CS-CDPD) technology. The effective range of applications for the CDPD network can be extended by using the CS-CDPD technology. To best understand CS-CDPD, let us examine the network layer reference model as shown in Figure 12.2. The three new functional elements are as follows:

- *The circuit-switched M-ES (CM-ES).* This is a variation of the M-ES with similar functionality.
- *The circuit-switched MD-IS (CMD-IS).* This is a variation of the MD-IS with similar functionality.
- *The circuit-switched modem bank (CS-MB).* This represents a modem bank developed to interoperate in a CS CDPD environment. The network interface between the CS-MB and the CMD-IS represents a LAN/WAN connection.

The interfaces CS and LW are new interfaces defined by the CS CDPD implementor guidelines. The external interface (reference point E) and the interservice provider interface (reference point I) are the same as the protocols and profiles specified in the CDPD System Specification v1.1. The use of the same E and I interfaces allows the CM-ES to switch between CS CDPD and CDPD service. The CM-ES can provide mobile IP/CLNP service to the subscriber. The standard CDPD features of authentication and encryption can be supported. The CM-ES gets access to the CDPD 1.1 network through either AMPS cellular or PSTN circuit-switched network facilities.

The CS CDPD service makes use of existing modem technology in the Circuit Switched CDPD network. This modem technology is used in the Modem Bank, the modems used by the Modem Bank and the modem used by the CM-ES.

12.3.1 CS CDPD Network Architecture

Figure 12.3 graphically describes the functional elements of the CS CDPD network.

The CS-MB to CMD-IS interface is within a single service provider and therefore is not a part of the I interface. This can be implemented in a variety of forms

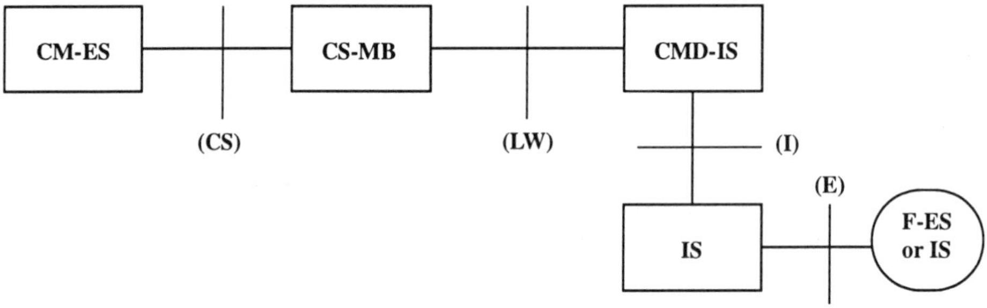

FIGURE 12.2 Network layer reference model for CS CDPD.

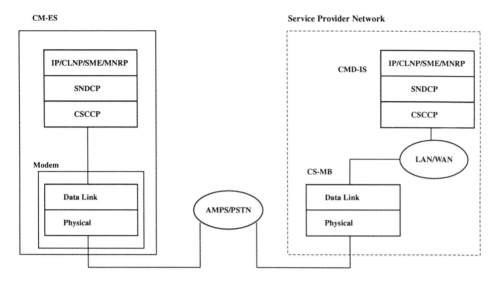

FIGURE 12.3 CS CDPD network architecture.

including Hayes AT and TCP/IP network components. As of the writing of this chapter, at least one service provider (GTE Mobilnet) has chosen to implement this interface using the TCP/IP-based protocol that is specified in the [1].

SNDCP and higher protocol functions in the CM-ES and the CMD-IS maintain the key functions of the CDPD network. A new protocol layer, the circuit-switched CDPD control protocol (CSCCP), provides additional peer-to-peer functions between the CM-ES and the CMD-IS. CSCCP provides a lightweight data link protocol with PDU framing, error detection, and frame loss detection. This layer is analogous to the MDLP layer in CDPD 1.1. The SNDCP/CSCCP interface is defined by the same primitives as the SNDCP/MDLP primitives.

The circuit-switched CDPD service can be accessed through any circuit-switched connection (landline or AMPS) at the option of the CM-ES. If the CM-ES is also a CDPD M-ES, the mobile can select the type of wireless CDPD service to use. The procedures for CDPD service selection are vendor specific and may use a variety of criteria. Some of these criteria are as follows:

- No CDPD 1.1 service or bad coverage available in the current service area.
- Poor throughput times for data transfers in a CDPD 1.1 environment. This can be due to the fact that excessive channel hopping is occurring and the availability of the channel for CDPD is low.
- Application program on the device. As an example, this can be large volumes of data transfers in the current application running on the subscriber device. Database-oriented data files fall under this category. A CDPD 1.1 channel is not optimal for this kind of data transfer.

12.4   EQUAL ACCESS PROVISION

The public-switched telephone network allows subscribers to choose their transport provider (loosely referred to as the long-distance service provider). In the data world, a similar analogy would be the transporter or other service provider of IP or CLNP packet service used to carry customer traffic. These are known as interexchange service providers (IXSP). The term "equal access" is a provision of access by the cellular service provider on an equal basis to all interexchange carriers, with respect to inter-LATA traffic originating within a LATA that has a destination outside of that cellular service provider's service area. Some of the CDPD service providers must provide equal access. This implies that the home MD-IS must be able to notify the serving MD-IS of the preferred IXSP selection. The preferred IXSP is available at the home MD-IS and is created as part of the provisioning database or subscriber profile at the time of customer subscription to CDPD service. The notification of the preferred interexchange service provider is done at the time of registration of the subscriber on the serving MD-IS by inclusion in the HomeInfo field.

The serving MD-IS can choose a variety of ways to implement equal access. These can range from the simple one port per IXSP to a more elegant encapsulation-based scheme [2]. The encapsulation schemes require the routers connected to the MD-IS to also understand the particular encapsulation protocol in use. This in some ways goes against the spirit in which the technology of CDPD evolved, wherein an emphasis was placed on making use of commercial, off-the-shelf routers. Typical routers in the market today do not support such encapsulation schemes.

Figure 12.4 illustrates one possible implementation of equal access at the serving MD-IS.

12.5   INTEROPERABILITY WITHIN WIRELESS NETWORKS

With increasing deployment of CDPD infrastructure to cover additional metropolitan areas, potential wireless data users have yet another new technology to consider before making a choice that could involve significant capital investment. The technologies are still evolving, and key performance, reliability, and coverage features have not been widely experimented, so users are reluctant to commit to a particular technology. Further, CDPD and other packet data networks such as ARDIS and RAM Data have proprietary air interfaces and network-specific protocol stacks, rendering the networks incompatible.

Interoperability between the different packet networks and application portability are two key issues that need to be resolved, which will assist users in adopting wireless data technology readily without having to risking investment in a specific technology. Further, interoperability will allow the networks to complement each other in the aspect of coverage, providing the users a wider geographical area of coverage than that provided by any single network.

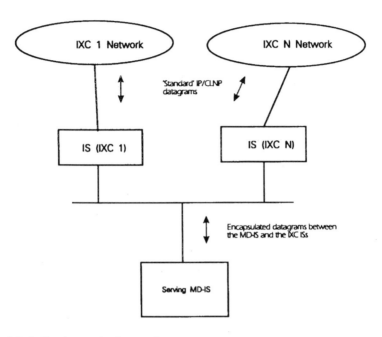

FIGURE 12.4 Equal access implementation.

Efforts in this direction are being addressed by a group called Personal Computer and Communication Association (PCCA) formed by ARDIS, Ericsson (associated with RAM Data), and McCaw (CDPD infrastructure provider, now part of AT&T).

The PCCA has been working on two fronts that will allow interoperability between the CDPD, ARDIS, and RAM Data nets: (1) wireless extensions that would allow an application with multiple networks, and (2) a common interface for IP to wireless network drivers that would allow routing among different networks.

Radiomail, whose architecture is shown in Figure 12.5, already provides routing between different nets. In this case, however, the implementation involves application-specific gateway implementation, which features address mapping and message-level protocol conversion at each network interface.

The PCCA's working strategy is outlined in the overall internetworking scheme shown in Figure 12.6.

The approach involves developing standard extensions to low-level drivers such as:

- Novell's Open Data Interface (ODI);
- Microsoft's Network Driver Interface Specifications (NDIS);
- Crynwyr Software's TCP/IP drivers for DOS.

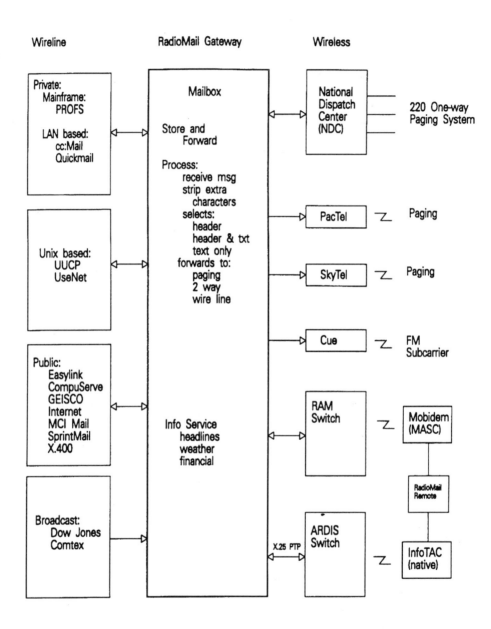

FIGURE 12.5 Radiomail architecture. (*Source:* [3].)

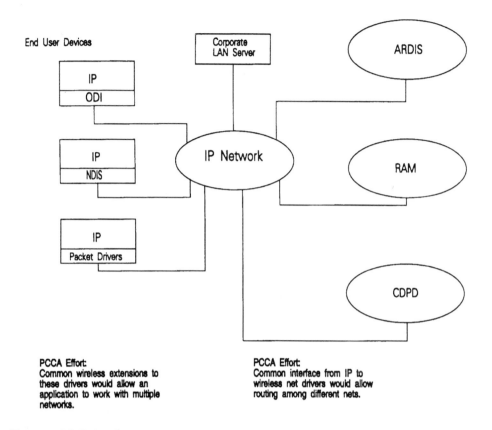

FIGURE 12.6 Wireless internetworking scheme. (*Source:* [4].)

Similarly, wireless driver extensions are being designed to allow access to any type of modem by untying the application software from the underlying hardware. This facility will nevertheless require multiple modems to access the different networks built as part of the hardware platform.

Application software also would require standardized APIs to spur application development. Ameritech has set up a Wireless Data Solutions Center that provides in a single lab commercial wireless products and infrastructure accessibility to the wireless networks and online databases [5]. To assist in developing, selling, evaluating, implementing, or integrating wireless software and hardware the Solution Center provides the following:

- Application development software;
- Circuit-switched modems;
- CDPD modems, boards, and phones;
- PDAs, organizers, handhelds, palmtop PCs, and laptops;

- Point of sale equipment;
- ATM hardware;
- Web browsers;
- Field service user application software from different vendors.

It is likely that other carriers will set up similar facilities to assist the growth in the use of wireless data networks. Carriers who are enjoying the rich revenue from the cellular voice market are hoping that the wireless data market will soon do the same.

12.6 PRICING STRUCTURES

Pricing involves various aspects of the wireless data network. In the PSTN arena where the operation is on a call-by-call basis, the accounting is based on the call time duration and distance. A telephone call is established over a router and that route does not change during that call. Distance is calculated based on the telephone numbers of the calling and called parties.

In the case of a wireless data network, like CDPD, the network uses standard IP and CLNP packet services. These are connectionless (i.e., there is no concept of a call at the service provider network). Customer data packets are transmitted independently and usually do not follow a set path. All the routing is done based on the addresses of the originator and destination. Accounting data in this case includes packet and octet (character) counts. The many different routes between the sending and receiving party and the fact that the packets may not reach their destination due to network hops, routing limitations, and outages complicates the accounting issues.

The vast amount of experience gained in pricing structures for the cellular world is, of course, useful, but needs modification in order to be applicable to the data world. Different service providers have adopted varying price structures for the CDPD service, but almost all of them have one aspect in common—the amount of data transferred in a unit of time like a month. The quantity of data is measured in simple octets or in multiples of 1 KB of data. Of course, marketing personnel at the service providers always have their role in trying out attractive pricing plans based on number of subscriber devices, type of data traffic application that the device is being used for, and the cost effectiveness of the device.

12.7 SECURITY ISSUES IN CDPD NETWORKS

The security issues from the CDPD subscriber perspective have been treated in Chapter 6. This section outlines the drawbacks in the authentication protocol and also attempts to discuss the security for the network in general and the infrastructure in particular. In view of the published drawbacks to the current authentication protocol, it is likely that the CDPD authentication protocol will be updated in the future to make the CDPD security provisions more robust.

12.7.1 Security Issues Related to the CDPD Authentication Protocol

This section summarizes the results of a study performed and presented to the CDPD forum as detailed in [6].

12.7.1.1 Current Authentication Protocol Summary

A summary of the message exchanges that are relevant to the authentication procedure is shown in Figure 12.7.

The important elements in the authentication protocol related to understanding possible threats to CDPD security are listed below:

- A Diffie-Hellman key exchange message establishes private encryption between the M-ES and its serving MD-IS. Here, security-critical information is being exchanged between two parties who do not know each other.
- The M-ES sends the authentication request with ESH (End System Hello) and RDR (Redirect Request) messages that transport the current credentials, the NEI and the shared historical record (SHR), which is essentially the ASN (authentication sequence number – 16 bits) and the ARN (authentication random number – 64 bits) to the home MDIS. The ESH packet is encrypted under the RC-4 cipher while the RDR is not.
- The home MDIS authenticates the M-ES and reports the confirmation or denial of the authentication back to the M-ES via the RDC (Redirect Confirm) and ISC (Intermediate System Confirm) packets. The home MDIS generates a new SHR, which is a new ARN (optionally and in typical implementations, it always generates a new ARN). The ASN is incremented by 1. In this sequence, the RDC is not encrypted but the ISC is.

In summary, the authentication information over the airlink is encrypted, but the information flow between the serving and the home MDISs are neither encrypted nor cryptographically authenticated.

12.7.1.2 Possible Attacks

Two potential areas of vulnerability in the authentication protocol can be identified.

Airlink Attack Since there is no authentication between an M-ES and its serving MDIS, an intruder in the airlink masquerading as a legitimate MD-IS can trick the M-ES into believing that it is a valid MD-IS. The M-ES will agree to an encryption key and will disclose its credentials (NEI and the current SHR) when it sends the

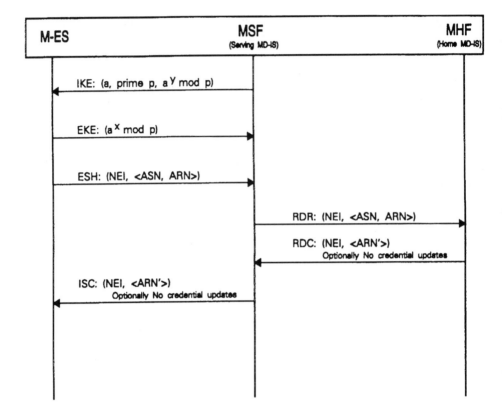

FIGURE 12.7 Current M-ES authentication protocol.

ESH packet to the fraudulent MD-IS. This form of attack is easy to accomplish as it only requires a functioning MDBS that can overpower the legitimate MDBS. Once the SHR is available, the attacker can use the known credentials to obtain services free of charge from the CDPD network and prevent the valid M-ES from obtaining services from the CDPD network.

Wireline-Side Attack Across the I interfaces in the wired CDPD backbone, M-ES authentication messages flow unencrypted between MD-ISs. An attacker who monitors the RDR messages can easily obtain the current credentials. This information can be intercepted or modified to deny services to legitimate M-ESs.

In addition, the unencrypted management messages such as RDF (Redirect Flush) and others, can provide ammunition to an intruder to cause service denial to a legitimate M-ES. Examples are as follows:

- A fraudulent MD-IS can send an RDF (Redirect Flush) message to a legitimate serving MDIS, which will terminate the data transfer services to the associated M-ES. Similarly, an RDE (Redirect Expiry) PDU sent from a fraudulent serving MDIS to the legitimate home MDIS will result in the home MDIS terminating the relaying of forward packets to the associated M-ES.
- An active attacker can generate a RDQ (Redirect Query) message to a legitimate serving MDIS, which will in turn initiate a new registration sequence from the M-ES. The resulting RDC (Redirect Confirm) message can be modified or a new one created at the appropriate time to provide a corrupted SHR to the M-ES that will lead to service denials during next registration attempt.

Although many of the above issues have not yet become critical, they soon will be when the CDPD customer base increases and security-sensitive applications become widespread [7,8].

12.7.2 General Infrastructure-Related Security

The CDPD network infrastructure, since it is IP- and CLNP-based and since it may be providing routing services for subscribers to get to the Internet world, is susceptible to unauthorized access. Misuse of the CDPD network can range from malicious hacking to repeated pings originating from outside the CDPD network, directed at any of the elements of the network. Potentially, these actions can allow data on the MD-IS to be compromised or can increase the operating cost of the target M-ES due to unnecessary traffic, in addition to an assault on the channel bandwidth.

The security of the infrastructure elements (namely, MDIS, MDBSs, NMS) can be ensured to a large degree by restricting access to the CDPD network. Access restrictions can be achieved by limited route access to the systems. There are primarily two methods available for limiting routes of the systems. First, the border router may be configured to filter traffic entering or exiting a service provider's network. The second method is to restrict the routes available to the border routers. However, either of these methods will limit the flexibility of the CDPD user to communicate with systems from all areas of the worldwide network. The subscriber should be able to decide if his M-ES's NEI will be advertised to the outside world. This flexibility does involve a great overhead at the border routers, especially with the implementation of CIDR for IP. The border router will either have to advertise the entire CIDR block and filter for specific numbers that do not wish to have access to the outside world or break up the CIDR block and represent blocks that want to be advertised to the outside world.

Unauthorized network usage can be accomplished at the External (E) and Inter-Service Provider (I) interface allowing for inter- and intra-CDPD Network access restriction points. This is accomplished by allowing only designated networks to peer with external ISs. Restrictions are to be planned out carefully so that aggregations are not broken based on access restrictions of sub-networks/nodes within an aggregation block.

References

[1]   Circuit Switched Cellular Digital Packet Data, Implementor Guidelines, Release 1.5, June 13, 1995, by the CDPD Forum.

[2]   RFC 1701: Hanks, S., T. Li, and D. Farinacci. Generic Routing Encapsulation, Oct. 1994.

[3]   DeRose, J. F., *The Wireless Data Handbook*, Quantum Publishing, 1994.

[4]   Wexler, J., "Wireless rivals team to ease interoperability," *Network World*, Nov. 14, 1994.

[5]   Hall, J,. "Ameritech unleashes a new resource for the expanding wireless market," *Wireless*, Vol. 4, No. 5, Sept./Oct. 1995.

[6]   Frankel, Y., A. Herzberg, P. A. Karger, H. Krawczyk, C. A. Kunzinger, and M. Yung, "Security Issues in a CDPD Wireless Network," *IEEE Personal Communications*, Vol. 2, No. 4, Aug. 1995, pp. 16–27.

[7]   Frankel, Y., et al., CDPD Fraud Prevention and Availability, TR-0296-12-94-589, GTE Laboratories, Inc., Waltham, MA, Dec. 1, 1994.

[8]   Molva, R., D. Samfat, and G. Tsudik, "Authentication of Mobile Users," *IEEE Network*, Special Issue on Mobile Communications, March/April 1994, pp. 26–34.

APPENDIX A
▼▼▼

DIRECTORY OF COMPANIES AND ORGANIZATIONS RELEVANT TO CDPD

ORGANIZATIONS, FORUMS

CDPD Forum
401 North Michigan Avenue, Suite 2200
Chicago, IL 60611
Tel: (800) 335–2373

Cellular Telecommunications Industry Association (CTIA)
1133 21st Street NW
Washington, DC 20036
Tel: (202) 785–0081

Federal Communications Commission (FCC)
1919 M Street NW, Room 814
Washington, DC 20554
Tel: (202) 632–6600

Infrared Data Association (IrDA)
P.O. Box 495
Brookdale, CA 95007
Tel: (408) 338–0924

Portable Computer and Communications Association (PCCA)
P.O. Box 924
Brookdale, CA 95007
Tel: (408) 338–0924

Telcommunications Industry Association (TIA)
2001 Pennsylvania Avenue NW, Suite 800
Washington, DC 20006
Tel: (202) 457–4912

MODEM VENDORS

Cincinnati Microwave
One Microwave Plaza
Cincinnati, OH 45249-9502
Tel: (513) 489–5400

IBM
P.O.Box 12195, Department K82A
200 Silicon Drive
Research Triangle Park, NC 27709
Tel: (919) 254–5826

Motorola
1303, East Algonquin Road
Schaumburg, IL 60196
Tel: (708) 576–5000

Pacific Communication Sciences, Inc. (PCSI)
10075 Barnes Canyon Road
San Diego, CA 92121
Tel: (619) 535–9500

Sierra Wireless
13151 Vanier Place, Suite 260
Richmond, British Columbia V6V 2J2, Canada
Tel: (604) 231–1114

Sony Electronics, Inc.
16450, West Bernado Drive
San Diego, CA 92127
Tel: (619) 673–2962

APPLICATION DEVELOPERS

AirLink Communications
4340 Stevens Creek, Suite 240
San Jose, CA 95129
Tel: (408) 261–6601

Mobileware Corporation
2425 North Central Expressway, No 1001
Richardson, TX 75080
Tel: (214) 952–1271

Performance Computing, Inc (PCI)
6300 Wilson Lane
Bethesda, MD 20817
Tel: (301) 320–6344
Fax: (301) 229–1020

Racotek
7301 Ohms Lane, Suite 200
Minneapolis, MN 55439
Tel: (612) 832–9800

Radiomail Corp
2600 Campus Drive
San Mateo, CA 94403
Tel: (415) 572–6001

SEI/Information Delivery Services
212 East Ohio Street, 2nd Floor
Chicago, IL 60611-3203
Tel: (312) 440–8383

Software Corporation of America, Inc.
100 Prospect Street
Stamford, CA 06901
Tel: (203) 359–2773

TEKnique, Inc.
911 North Plum Grove Road
Schaumburg, IL 60173
Tel: (708) 706–9700

Wireless Connect
2177 Augusta Place
Santa Clara, CA 95051
Tel: (408) 296–1546

CARRIERS

AT&T
67 Whippany Road, Room 2A327
Whippany, NJ 07981
Tel: (201) 386–2694

Alltel Mobile Communications
10825 Financial Parkway, No 401
Little Rock, AR 72211

Ameritech Cellular
2000 West Ameritech Center Drive
Hoffman Estates, IL 60195-5000
Tel: (708) 765–3900

Bell Atlantic Mobile Systems
180 Washington Valley Road
Bedminister, NJ 07921
Tel: (908) 306–7666

Bellsouth Cellular Corporation
1100 Peachtree Street Northeast, Suite 1
Atlanta, GA 30309
Tel: (404) 713–0684

GTE
245 Perimeter Center Parkway
Atlanta, GA 30346
Tel: (404) 804–3450

McCaw Cellular Communications
10230 Northeast Points Drive
Kirkland, WA 98033
Tel: (206) 803–4617

MCI Telecommunications
1801 Pennsylvania Avenue Northwest
Washington, DC 20006
Tel: (202) 887–2305

Mobility Canada
20 Carlson Court
Etobocoke, Ontario M9W 6V4, Canada
Tel: (416) 213–3179

NYNEX Mobile Communications Company
2000 Corporate Drive
Orangeburg, NY 10962

Southwestern Bell Mobile Systems, Inc.
17330 Preston Road, Suite 100A
Dallas, TX 75252
Tel: (214) 733–2169

Sprint Cellular
8725 Higgins Road
Chicago, IL 60631
Tel: (312) 399–2461

OTHER WIRELESS COMPANIES

Advanced Network and Services, Inc.
100, Phoenix Drive
Ann Arbor, MI 48108-2202

Air Communications
274 San Geromino Way
Sunnyvale, CA 94086
Tel: (408) 749–9883

AirTouch Cellular
1340 Treat Boulevard, Suite 500
Walnut Creek, CA 94596
Tel: (510) 988–4422

APEX Data, Inc.
6624 Owens Drive
Pleasanton, CA 94588-3334
Tel: (510) 416–5656

ARDIS
300 Knightbridge Parkway
Lincolnshire, IL 60069
Tel: (708) 913–4233

CellPort Labs, Inc.
4730 Walnut Street
Boulder, CO 80301
Tel: (303) 541–0722

Cue Paging Corporation
2737 Campus Drive
Irvine, CA 92715
Tel: (714) 752–9200

EMBARC Communications Service
1500 Northwest 22nd Street
Boynton Beach, FL 33426
Tel: (407) 364–2519

Ericsson Radio Systems
740 East Campbell Road
Richardson, TX 75081
Tel: (214) 238–3299

General Magic
2465 Latham Street
Mountain View, CA 94040
Tel: (415) 965–0400

Geotek Communications
20 Craig Road
Montvale, NJ 07445
Tel: (201) 930–9305

GeoWorks
960 Atlantic Avenue
Alameda, CA 94501
Tel: (510) 814–1660

Hughes Network Systems
11717 Exploration Lane
Germantown, MD 20876
Tel: (301) 428–5500

Iridium, Inc.
1401 H Street NW
Washington DC 20005
Tel: (202) 326–5600

Metricom
980 University Avenue
Los Gatos, CA 95030
Tel: (408) 399–8200

MobileComm
1800 East County Line Road
Ridgeland, MS 39157
Tel: (601 977–0888

Nationwide Wireless Network
200 S. Lamar Street
Jackson, MS 39201
Tel: (601) 944–7209

Nextel
201 Route 17 North
Rutherford, NJ 07070
Tel: (201) 438–1400

Nomadic Systems, Inc.
300 Ferguson Drive, Suite 200
Mountain View, CA 94043
Tel: (415) 335–4310

Omnipoint Corporation
7150 Campus Drive
Colorado Springs, CO 80920
Tel: (719) 548–1200

Paging Network, Inc.
4965 Preston Park Boulevard, Suite 600
Plano, TX 75093
Tel: (214) 985–4100

Palm Computing
4410 El Camino Real, Suite 108
Los Altos, CA 94022
Tel: (415) 949–9560

PenWare
845 Page Mill Road
Palo Alto, CA 94304
Tel: (415) 858–4920

Proxim, Inc.
295 North Bernado Avenue
Mountain View, CA 94043
Tel: (415) 960–1630

Qualcomm, Inc.
6455 Lusk Boulevard
San Diego, CA 92121
Tel: (619) 587–1121

RAM Mobile Data
3 University Plaza, Suite 600
Hackensack, NJ 07601
Tel: (201) 343–9400

SkyTel
1350 I Street NE, Suite 1100
Washington, DC 20005
Tel: (202) 408-7444

Telular
920 Deerfield Parkway
Buffalo Grove, IL 60089
Tel: (708) 256-8000

Tetherless Access
43730 Vista Del Mar
Fremont, CA 94539
Tel: (510) 659-0809

Trimble Navigation
645 North Mary Avenue
Sunnyvale, CA 94088
Tel: (408) 730-2900

Wireless Access
125 Nicholson Lane
San Jose, CA 95134
Tel: (408) 383-1900

APPENDIX B

▼▼▼

State of CDPD Deployment

Metro Area	Coverage	Availability	Carrier	Integrator
New York, NY	Initial Core	Limited Full	AT&T Wireless Services Bell Atlantic NYNEX	AT&T Wireless Services AT&T
Chicago, IL	Full	Full	Ameritech	Hughes
Philadelphia, PA	Full Full	Full Full	Bell Atlantic NYNEX Comcast	AT&T AT&T
Detroit-Ann Arbor, MI	Full	Limited	Ameritech	Hughes
San Fransisco-Oakland, CA	Full	Full	GTE Mobilnet	Hughes
Washington, D.C	Full	Full	Bell Atlantic NYNEX	AT&T
Dallas, TX	Core	Limited	AT&T Wireless Services	AT&T Wireless Services
Houston, TX	Core	Full	GTE Mobilnet	Hughes
Miami, FL	Full	Full	AT&T Wireless Services	AT&T Wireless Services
Pittsburgh, PA	Full Initial	Full Limited	Bell Atlantic NYNEX AT&T Wireless Services	Hughes AT&T Wireless Services
Baltimore, MD	Full	Full	Bell Atlantic NYNEX	AT&T
Minneapolis-St.Paul, MN	Full	Full	AT&T Wireless Services	AT&T Wireless Services
Cleveland, OH	Core	Full	GTE Mobilnet	Hughes
Seattle-Everet, WA	Core	Full	AT&T Wireless Services	AT&T Wireless Services
Tampa-St.Petersburg, FL	Core Core	Full Full	GTE Mobilnet AT&T Wireless Services	Hughes AT&T Wireless Services
Phoenix, AZ	Full	Full	Bell Atlantic NYNEX	Hughes
San Jose, CA	Full	Full	GTE Mobilnet	Hughes
Portland, OR	Full	Full	AT&T Wireless Services	AT&T Wireless Services
Hartford, CT	Full Full	Full Full	Bell Atlantic NYNEX SNET Mobility	Hughes AT&T
Salt Lake City/Ogden, UT	Full	Full	AT&T Wireless Services	AT&T Wireless Services
Bridgeport-Stamford Norwalk-Danbury, CT	Full Full	Full Full	Bell Atlantic NYNEX SNET Mobility	Hughes AT&T

Metro Area	Current Coverage	Availability	Carrier	Integrator
Norfolk, VA Beach-Portsmouth, VA/NC	Core	Full	GTE Mobilnet/Contel	Hughes
Oklahoma City, OK	Full	Full	AT&T Wireless Services	AT&T Wireless Services
New Haven, CT	Full / Full	Full / Full	Bell Atlantic NYNEX / SNET Mobility	Hughes / AT&T
Akron, OH	Core	Full	GTE Mobilnet	Hughes
Gary, IN	Full	Full	Ameritech	Hughes
Tulsa, OK	Full	Full	AT&T Wireless Services	AT&T Wireless Services
Allentown, VA	Initial	Full	Bell Atlantic NYNEX	AT&T
Richmond, VA	Core	Full	GTE Mobilnet	Hughes
Charlotte, NC	Core	Full	Bell Atlantic NYNEX	Hughes
New Brunswick, NJ	Core / Initial	Full / Full	Bell Atlantic NYNEX / Comcast	AT&T / AT&T
Wilmington, DE	Core / Core	Full / Full	Bell Atlantic NYNEX / Comcast	AT&T / AT&T
Atlantic City, NJ	Initial	Full	Comcast	AT&T
Trenton/Princeton, NJ	Core	Full	Comcast	AT&T
Long Branch-Asbury Park, NJ	Initial	Full	Bell Atlantic NYNEX	AT&T
West Palm Beach-Boca Raton, FL	Full	Full	AT&T Wireless Services	AT&T Wireless Services
Tucson, AZ	Core	Full	Bell Atlantic NYNEX	Hughes
Las Vegas, NV	Core	Full	AT&T Wireless Services	AT&T Wireless Services
New London-Norwich, CT	Full	Full	Bell Atlantic NYNEX	Hughes
Fredrick, MD-3	Initial	Full	Bell Atlantic NYNEX	AT&T
Hunterdon, NJ-1	Initial	Full	Bell Atlantic NYNEX	AT&T
Farmington, NM	Full	Full	Sprint	Motorola
Raleigh-Durham, NC	Core	Full	GTE Mobilnet	NTI

Metro Area	Coverage	Availability	Carrier	Integrator
New York-Nassau-Suffolk, NY	Core	Full	AT&T Wireless Services	AT&T Wireless Services
Newark, Jersey City, Paterson-Clifton-Pasaic, NJ	Core	Full	Bell Atlantic NYNEX	AT&T
Los Angeles-Longbeach/Anaheim Santa Ana-Garden Grove/Riverside San Bernadino-Ontario, CA			AirTouch BellSouth/AT&T Wireless Services	
Boston-Lowell-Brockton-Lawrence Haverhill, MA/NH		Trial	Bell Atlantic NYNEX	
St.Louis, MO-IL	Core	Trial	Ameritech	AT&T
Miami-Ft.Lauderdale-Hollywood, FL	Full	Full	BellSouth AT&T Wireless Services	AT&T Wireless Services
Atlanta, GA		Trial	BellSouth	AT&T
San Diego, CA	Initial	Trial	GTE Mobilnet	
Denver-Boulder, CO		Full	AT&T Wireless Services	AT&T Wireless Services
Milwaukee, WI	Initial	Trial	Ameritech BellSouth	
Cincinnati, OH-KY-IN	Initial	Trial Trial	AirTouch/CCI Ameritech	
Kansas City, MO-KS		Trial	AirTouch/AT&T Wireless Services	
Indianapolis, IN	Core	Trial	BellSouth GTE Mobilnet	Ericsson Hughes
New Orleans, LA			BellSouth	
Columbus, OH	Initial	Trial	AirTouch/CCI Ameritech	AirTouch
Hartford-New Britain-Bristol, CT	Full Full	Full Full	Bell Atlantic NYNEX SNET Mobility	Hughes AT&T
San Antonio, TX		Full	AT&T Wireless Services	AT&T Wireless Services
Sacramento, CA		Trial	AT&T Wireless Services	AT&T Wireless Services
Portland, OR	Full	Full	AT&T Wireless Services	AT&T Wireless Services

Metro Area	Coverage	Availability	Carrier	Integrator
Memphis, TN-AR-MS		Trial	BellSouth GTE Mobilnet	Hughes
Louisville, KY		Trial	GTE Mobilnet	Hughes
Providence-Warwick-Pawtucket, RI			Bell Atlantic NYNEX	
Birmingham, AL		Trial	BellSouth GTE Mobilnet	Hughes
Nashville-Davidson, TN	Core	Trial	BellSouth GTE Mobilnet	Hughes
Greensboro-Winston-Salem-Highpoint, NC	Initial	Trial	GTE Mobilnet Sprint	NTI Motorola
New Haven-West Haven-Waterbury Meriden, CT	Full Core	Full Limited	BellAtlantic NYNEX SNET Mobility	Hughes AT&T
Honolulu, Hawaii		Trial	BellSouth GTE Mobilnet	Hughes
Jacksonville, FL		Full Full	AT&T Wireless Services BellSouth	AT&T Wireless Services
Gary-Hammond-East Chicago, IN	Full	Full	Ameritech	Hughes
Allentown-Bethlehem-Easton, PA-NJ	Initial	Full	Bell Atlantic NYNEX	AT&T
Orlando, FL		Full	AT&T Wireless Services BellSouth	AT&T Wireless Services
Charlotte-Gastonia, NC	Core	Full	Bell Atlantic NYNEX	Hughes
New Brunswick-Perth-Amboy Sayreville, NJ	Core Core	Full Full	Bell Atlantic NYNEX Comcast	AT&T AT&T
Wilmington, DE-NJ-MD	Core Full	Full Full	Bell Atlantic NYNEX Comcast	AT&T AT&T
Long Branch-Ashbury Park, NJ	Initial	Full	Bell Atlantic NYNEX	AT&T
West Palm Beach-Boca Raton, FL	Full	Full	AT&T Wireless Services	AT&T Wireless Services

Metro Area	Coverage	Availability	Carrier	Integrator
Fresno, CA	Core	Trial	GTE Mobilnet	Hughes
Austin, TX	Core	Trial	GTE Mobilnet	Hughes
El Paso, TX		Trial	Bell Atlantic NYNEX	
Albuquerque, NM		Trial	Bell Atlantic NYNEX	
Las Vegas, NV	Core Initial	Full Trial	AT&T Wireless Services Sprint	AT&T Wireless Services Motorola
Bakersfield, CA	Core	Trial	GTE Mobilnet	
New London-Norwich, CT	Full	Full	BellAtlantic NYNEX	Hughes
Las Cruces, NM		Trial	BellAtlantic NYNEX	
Arizona 2-Coconino		Trial	Bell Atlantic NYNEX	AT&T Wireless Services
Arizona 5 - Gila		Trial	Bell Atlantic Nynex	Hughes
Maryland 3- Fredrick	Initial	Full	Bell Atlantic NYNEX	AT&T
New Jersey 1 - Hunterdon	Initial	Full	Bell Atlantic NYNEX	AT&T
New Mexico 1 - Farmington	Full	Full	Sprint	Motorola

▼▼▼

ABOUT THE AUTHORS

Dr. Muthuthamby Sreetharan was born in Jaffna (Thamil Eelam) Sri Lanka. He obtained his B.S. with highest honors from University of Sri Lanka and taught electrical engineering for two years. He studied at University of Manchester Institute of Science and Technology and held a research fellowship at Brunel University's (UK) parallel processing group, obtaining his M.S. in digital electronics and Ph.D. in 1979 and 1982, respectively. He is the president of Performance Computing, Inc. (PCI), which specializes in embedded systems development for voice/data and mobile communication systems. Prior to his involvement in wireless networks, he had worked on developing several data communication systems and digital control systems. Currently he is a consultant with Hughes Network Systems, involved in the continuing deployment and enhancement of the CDPD system. He can be reached via electronic mail at sree@pcinet.com.

Mr. Rajiv Kumar is currently working on developing personal air communication technology (pACT)–based systems at PCSI, San Diego. He is currently leading the development of a subscriber product that supports advanced messaging and two-way paging using the pACT suite of protocols. Prior to this, Rajiv was working at Hughes Network Systems as the lead system architect on their CDPD infrastructure offering. Rajiv has worked on designing and developing a variety of cellular systems, mainly working on the network components. Before he came into the cellular arena, he worked on local and wide area communications and switching systems. His education has been primarily in Indian Institute of Technology, Bombay, India.

▼▼▼

INDEX

The Artech House Telecommunications Library

Vinton G. Cerf, Series Editor

Writing Disaster Recovery Plans for Telecommunications Networks and LANs,
  Leo A. Wrobel

X Window System User's Guide, Uday O. Pabrai

For further information on these and other Artech House titles, contact:

Artech House
685 Canton Street
Norwood, MA 02062
617-769-9750
Fax: 617-769-6334
Telex: 951-659
e-mail: artech@artech-house.com

Artech House
Portland House, Stag Place
London SW1E 5XA England
+44 (0) 171-973-8077
Fax: +44 (0) 171-630-0166
Telex: 951-659
e-mail: artech-uk@artech-house.com